Advanced Micro-Level Experimental Techniques for Food Drying and Processing Applications

Advances in Drying Science and Technology

Series Editor Dr. Arun S. Mujumdar

It is well known that the unit operation of drying is a highly energy-intensive operation encountered in diverse industrial sectors ranging from agricultural processing, ceramics, chemicals, minerals processing, pulp and paper, pharmaceuticals, coal polymer, food, forest products industries as well as waste management. Drying also determines the quality of the final dried products. The need to make drying technologies sustainable and cost effective via application of modern scientific techniques is the goal of academic as well as industrial R&D activities around the world.

Drying is a truly multi- and interdisciplinary area. Over the last four decades the scientific and technical literature on drying has seen exponential growth. The continuously rising interest in this field is also evident from the success of numerous international conferences devoted to drying science and technology.

The establishment of this new series of books entitled Advances in Drying Science and Technology is designed to provide authoritative and critical reviews and monographs focusing on current developments as well as future needs. It is expected that books in this series will be valuable to academic researchers as well as industry personnel involved in any aspect of drying and dewatering.

The series will also encompass themes and topics closely associated with drying operations, e.g., mechanical dewatering, energy savings in drying, environmental aspects, life cycle analysis, technoeconomics of drying, electrotechnologies, control and safety aspects, and so on.

About the Series Editor

Dr. Arun S. Mujumdar is an internationally acclaimed expert in drying science and technologies. He is the Founding Chair in 1978 of the International Drying Symposium (IDS) series and Editor-in-Chief of Drying Technology: An International Journal since 1988. The 4th enhanced edition of his Handbook of Industrial Drying published by CRC Press has just appeared. He is recipient of numerous international awards including honorary doctorates from Lodz Technical University, Poland and University of Lyon, France.

Please visit www.arunmujumdar.com for further details.

Advanced Micro-Level Experimental Techniques for Food Drying and Processing Applications

Azharul Karim
Sabrina Fawzia
Mohammad Mahbubur Rahman

CRC Press is an imprint of the
Taylor & Francis Group, an **informa** business

First edition published 2022
by CRC Press

6000 Broken Sound Parkway NW, Suite 300, Boca Raton, FL 33487-2742

2 Park Square, Milton Park, Abingdon, Oxon, OX14 4RN

Library of Congress Cataloging-in-Publication Data
Names: Karim, Azharul, author. | Fawzia, Sabrina, author.
Title: Advanced micro-level experimental techniques for food drying and processing applications / Azharul Karim and Sabrina Fawzia.
Description: First edition. | Boca Raton : CRC Press, 2022. | Includes bibliographical references and index.
Identifiers: LCCN 2021027698 (print) | LCCN 2021027699 (ebook) | ISBN 9780367472160 (hardback) | ISBN 9780367496999 (paperback) | ISBN 9781003047018 (ebook)
Subjects: LCSH: Food industry and trade--Technological innovations. | Food--Drying--Equipment and supplies. | Food--Microstructure. | Microphotography. | Industrial microscopy. | Radiography, Industrial.
Classification: LCC TP371 .K37 2022 (print) | LCC TP371 (ebook) | DDC 664/.02--dc23
LC record available at https://lccn.loc.gov/2021027698
LC ebook record available at https://lccn.loc.gov/2021027699

ISBN: 978-0-367-47216-0 (hbk)
ISBN: 978-0-367-49699-9 (pbk)
ISBN: 978-1-003-04701-8 (ebk)

DOI: 10.1201/9781003047018

Typeset in Times
by MPS Limited

Contents

Preface

Food microstructure is defined as the spatial organisation of structural components and their interactions. Plant-based food materials are porous in nature and heterogeneous in structure, which undergo significant changes during food processing. Understanding these microstructural changes is important for improving the product quality, reducing the food waste and developing accurate mathematical models for the food processes. Many techniques have been developed recently for micro-level investigations of food material, and these advancements have opened opportunities for food scientists, researchers and industrial engineers to significantly improve the process performance and develop advanced models. However, currently there is no comprehensive reference book that presents the details of these recent developments and therefore the research community and industries are not fully aware of these techniques.

This book describes in detail the advanced micro-imaging methods applicable in food engineering research and industrial processes and therefore should be an important reference for food engineers, food scientists, food processing engineers, researchers and undergraduate and postgraduate students in this field. The readers are expected to:

- Learn and understand the working principles of the microimaging techniques used for the microstructure investigation of food products
- Learn the detailed procedure of applying these techniques in food processing
- Understand the current challenges of developing efficient and novel food processing systems and their links with micro-level investigations and microimaging

The book highlights the importance of understanding the microstructural changes of plant-based food materials using available techniques. Such methods have previously been unfeasible due to the lack of adequate experimental testing. This book presents most important techniques, together with their relative advantages and disadvantages, which will give the readers clear guidelines on the selection of right techniques for specific investigation to get maximum benefit.

Azharul Karim

Authors

Dr Azharul Karim is currently working as an Associate Professor in the school of Mechanical, Medical and Process Engineering, Queensland University of Technology, Australia. He received his PhD degree from Melbourne University in 2007. Dr Karim has authored over 200+ peer-reviewed articles, including 110 high quality journal papers, 13 peer-reviewed book chapters, and 4 books. His papers have attracted about 5,000+ citations with h-index 38. His research has a very high impact worldwide as demonstrated by his overall field weighted citation index (FWCI) of 2.10. He is an editor/board member of six reputed journals including *Drying Technology* and *Nature Scientific Reports* and a supervisor of 26 past and current PhD students. He has been a keynote/distinguished speaker at scores of international conferences and an invited/keynote speaker in seminars in many reputed universities worldwide. He has won multiple international awards for his outstanding contributions in multidisciplinary fields. His research is directed towards solving acute food industry problems by advanced multiscale and multiphase food drying models of the cellular water using theoretical/computational and experimental methodologies. He is the recipient of numerous national and international competitive grants amounting to $3.15 million.

Dr Sabrina Fawzia is currently working as a senior lecturer in civil engineering at Queensland University of Technology, Australia. She is a structural engineering expert, and research focuses on the development of the high performance structural members and structural strengthening/retrofitting by using fibre reinforced polymer (FRP) material technology. Through her scholarly, innovative, high-quality research, she has established her national and international standing. Dr Fawzia's excellence in research has been demonstrated by high-quality refereed publications (98 publications: 1,795 citations, h-index = 21 Google Scholar), two ARC LIEF grants ($1.9M), one international grant ($80K), five QUT internal grants ($79K), QUT's SEF Award for Excellence in postgraduate research supervision, 10 PhD's and 3 master's by research completions, being invited by reputed universities for seminars and the establishment of national and international collaborative research relationships. Her recent research interest includes the microstructural analysis of for structures.

Dr Mohammad Mahbubur Rahman received his PhD degree from Queensland University of Technology (QUT), Australia, in 2018. Currently, he is working as a visiting research fellow at the Queensland University of Technology (QUT). He received his BSc Degree in Electrical and Electronic Engineering (EEE) from the Chittagong University of Engineering and Technology (CUET), Bangladesh, in 2010 and Master of Engineering Science from the University of Malaya (UM), Malaysia, in 2014. His research interest includes mathematical modelling, drying process optimization and renewable energy.

1 Prospect of Micro-level Investigations in Food Processing Applications

1.1 INTRODUCTION

Food processing industry is one of the largest manufacturing industries worldwide and employs a large proportion of the workforce [1]. Drying is one of the major processes in the food processing industry and a dominant food preservation method [2]. However, the current thermal processing systems of food, such as drying, are highly energy-intensive and lengthy processes, which result in significant food quality deterioration [3]. These problems could not be resolved as the thermal systems are designed and process parameters are determined based on a limited empirical knowledge, without considering the underlying micro-level dynamic changes during thermal processing [4]. Fruits and vegetables (FV) have a heterogeneous structure with hygroscopic characteristics and undergo significant modifications at different length scales during food processing due to simultaneous heat, mass and momentum transfer [5,6]. In reality, FV drying is a multilevel problem – spanning from microscale to dryer scale (Figure 1.1): (i) a complex transport problem – strong geometrical and material nonlinearities; (ii) a coupling problem – micro-macro-dryer; and a (iii) a heterogeneous deformation problem – samples undergo nonuniform morphological changes [7].

During thermal processing, conditions and properties change over time: both rapidly and gradually, transiently and permanently, in structure and in appearance, which not only impact drying kinetics but significantly impact morphological and quality attributes. These dynamic property and anisotropic morphological changes occur at a micro-level and critically govern the thermal processes and the product characteristics [3]. Generally, microstructural elements below 100 μm significantly influence the transport properties and the physical behaviour of food materials [8]. Quality deterioration during FV thermal processing is heavily linked with microstructural changes and its associated deformation [9,10]. However, the current knowledge about micro-level investigations has been limited. Understanding such microstructural changes has been previously unfeasible due to the lack of experimental methods for cellular-level investigations.

Recent advancement in micro-level investigation methods has opened opportunities for food scientists and engineers to give new insights into food processing problems. For example, a recent magnetic resonance imaging (MRI)–based study uncovered that during food drying, the intracellular water

DOI: 10.1201/9781003047018-1

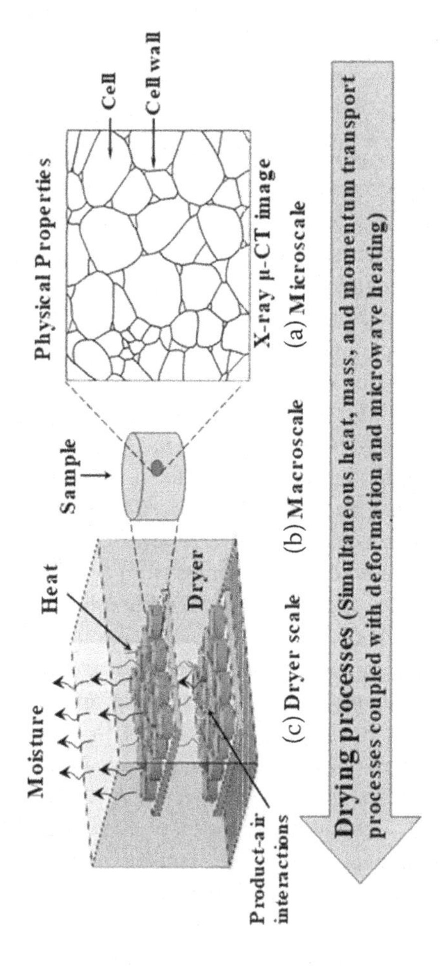

FIGURE 1.1 Understanding the different scales of food drying.

(ICW) migrates from intracellular spaces to intercellular spaces by progressively rupturing the cell walls [7]. This cell rupturing has a flow-on effect on transport, deformation and food quality.

Retaining the quality aspects of FV during drying is a major concern since the acceptability of the product depends on the overall quality, particularly the nutritional, colour, physical and sensory attributes. Significant changes in product quality attributes, including health-promoting components [11], occur during thermal processing due to the microstructural changes and simultaneous effects of heat and mass transfer that influence biochemical reactions. It is expected that more micro-level investigations using advanced methods can help resolve these problems.

1.2 RECENT DEVELOPMENTS IN MICRO-LEVEL INVESTIGATION METHODS

During thermal treatment, all the transport processes and associated morphological changes such as cell wall cracking or shrinking initiate at the cell level. Changes in food cellular structure during food processing significantly influence the nutritional, physical and sensory properties of products. Food processing performance is also influenced by the microstructural changes. For example, during the drying of foods, drying kinetics change due to the dynamic change in the microstructures. Therefore, to achieve the desired process performance and product quality, a better understanding of the food structure and its evolution during food processing is required. Understanding cell-level water transport processes and associated morphological and property changes is important. Moreover, food dying and other processing models have advanced from bulk-level to micro-level and from single-scale to multiscale models. Micro-level and multiscale mathematical models require the knowledge of microstructure and cellular properties. More advanced models may even need the knowledge about three-dimensional (3D) microstructure. The lack of knowledge about the recent advances in micro/cell-level investigation methods is one of the main barriers for making progress in developing multiscale models of food processing.

Many techniques have been developed for micro-level investigations in the recent time. All these techniques have relative advantages and disadvantages, and therefore selecting right techniques for specific investigation is paramount for ensuring desired outcomes. Some techniques, such as the light and electron microscopy, are considered invasive methods and therefore involve separate sample preparation for each test, while other methods such as MRI and atomic force microscopy (AFM) have specific applications. X-ray micro-computed tomography (μCT) is a relatively new imaging method that enables non-destructive and non-invasive microstructural investigations at resolutions higher than 1 mm, and the analysis aimed at the internal examination of the structural arrangement of products [11]. X-ray technology was invented in 1895, and this technology was advanced as X-ray computed tomography (CT) and introduced into clinical practices in the 1970s. The first application of X-ray μCT in food science was reported in 1991, when it was used to detect the maturity of green tomatoes. Besides being used in the millimetre to micron resolution range, the recent advancement of the technique has enabled sub-micrometre or nanometre (X-ray nano-CT) pixel resolution. The same sample can thus be scanned multiple times, which is particularly important food engineering research where microstructural changes over time need to be monitored. For example, properties of food materials dynamically change during drying and therefore same sample needs to be scanned at different stages of drying [12]. X-ray μCT, therefore, enables the same samples to be studied in their original state as well as at different stages of processing.

X-ray CT is found to have its primary applications in medical science, and other areas of science have reported wide range of applications of this technology. The success of this imaging method in these fields encourages its use in food science. In the food industry, there is a need for quantitative techniques that can accurately characterise food products, with the aim of establishing relationships between the microstructure and the food quality and controlling the process parameters accordingly. Three-dimensional visualisation of the food microstructure can significantly help understand these properties and establish their relationships with processing conditions [13]. This can also help design the products with desired attributes. Although evidence suggests that there is a relationship between the microstructure and the food quality, an established method for investigating that relationship is not available yet. There is an opportunity to develop such a method based on advanced micro-imaging techniques.

In recent years, consumers are becoming more concerned about their wellness and safety and are therefore becoming interested in foods with healthy nutrients and bioactive compounds. Similarly, sensory characteristics such as taste and aroma are very important from the consumer perspective. Therefore, robust and non-destructive analytical methods are required for effectively analysing the composition and measuring the nutritional and physicochemical properties and functionality of food matrices, which will help the development and production of high-quality, nutrition-rich and safe foods for future. Nuclear magnetic resonance (NMR) spectroscopy is one of the most powerful and versatile analytical techniques that can be applied to food materials, particularly in FV products.

Nuclear magnetic resonance is a rapid, reliable and non-invasive technology and can be used to detect the quality of food. However, due to the high cost of the equipment and maintenance, low detection limit and sensitivity, the requirement of highly skilled manpower and the safety issues due the magnetic field, so far, the practical applications are limited. Efforts are needed to promote the applications of this method for a broader range of foods. NMR was first used in food science to determine the moisture content and distribution of food, but with the recent improvements, NMR technology is applied to different food fields including quantitative analysis, conformational analysis, nutritional or functional aspects, quality detection and process control. The major advantages of NMR are that it is a non-destructive test and quantitative information can be retrieved [14]. However, the major limitations of this method are the detection limits and its sensitivity. NMR has been classified into two types based on the NMR spectrum and magnetic resonance techniques: (a) high-resolution NMR (HRNMR) and (b) low-field NMR (LF-NMR) [14].

Although many advances have been made recently in the development of technology to quantitatively explore micro-level structural and quality changes during processing, there is no comprehensive book that presents the details of these developments and therefore the research community and industries are not fully aware of these techniques. Global researchers, scientists and industrial engineers involved in the food processing need enough reference material on advanced techniques in micro-level investigations to understand the fundamental water transport processes and associated morphological changes during the thermal processing of the food material. Undergraduate and postgraduate students in food engineering, food science and other relevant fields also need a proper reference book on this topic. Although few books and chapters are available on microimaging methods, most of the methods were applied in other fields. There has been some advancement in applying these methods in food processing in the recent years. Therefore, a specific book that brings together these advancements will be of great help for these above-mentioned stakeholders. The three key features of the book are:

- This book provides details of the latest micro-level experimental techniques with a particular focus on microimaging techniques.
- It describes a detailed procedure of applying these techniques in food processing.
- It also describes the current challenges of developing efficient and novel food processing systems and their links with micro-level investigations and microimaging.

Most of the food materials have a heterogeneous structure with cellular diversity. Without a proper understanding of the cell-level transport process, quality changes and morphological changes, it is not possible to develop efficient food processing techniques such as food drying process. However, no book currently contains the latest developments in this area. This book fills this vacuum.

REFERENCES

1. Joardder, M.U., et al., *A micro-level investigation of the solid displacement method for porosity determination of dried food.* Journal of Food Engineering, 2015. **166**: pp. 156–164.
2. Mahiuddin, M., et al., *Development of fractional viscoelastic model for characterizing viscoelastic properties of food material during drying.* Food Bioscience, 2018. **23**: pp. 45–53.
3. Khan, M., et al., Investigation of cellular level of water in plant-based food material. In International Drying Symposium. 2016. Japan.

4. Rahman, M.M., M.U. Joardder, and A. Karim, *Non-destructive investigation of cellular level moisture distribution and morphological changes during drying of a plant-based food material*. Biosystems Engineering, 2018. **169**: pp. 126–138.
5. Abesinghe, A., et al., *Effects of ultrasound on the fermentation profile of fermented milk products incorporated with lactic acid bacteria*. International Dairy Journal, 2019. **90**: pp. 1–14.
6. Kumar, C., et al., *A porous media transport model for apple drying*. Biosystems Engineering, 2018. **176**: pp. 12–25.
7. Khan, M.I.H., et al., *Fundamental understanding of cellular water transport process in bio-food material during drying*. Scientific Reports, 2018. **8**(1): p. 15191.
8. Rahman, M.M., et al., *Multi-scale model of food drying: Current status and challenges*. Critical Reviews in Food Science and Nutrition, 2018. **58**(5): pp. 858–876.
9. Senadeera, W., et al., *Modeling dimensional shrinkage of shaped foods in fluidized bed drying*. Journal of Food Processing and Preservation, 2005. **29**(2): pp. 109–119.
10. Mulet, A., et al., *Effect of shape on potato and cauliflower shrinkage during drying*. Drying Technology, 2000. **18**(6): pp. 1201–1219.
11. Duc Pham, N., et al., *Quality of plant-based food materials and its prediction during intermittent drying*. Critical Reviews in Food Science and Nutrition, 2019. **59**(8): pp. 1197–1211.
12. Khan, M.I.H., et al., *Cellular level water distribution and its investigation techniques*. In: Chung-Lim L., Azharul K. (eds). Intermittent and Nonstationary Drying Technologies: Principles and Applications, 2017, CRC Press. pp. 193–210.
13. Pinzer, B., et al., *3D-characterization of three-phase systems using X-ray tomography: tracking the microstructural evolution in ice cream*. Soft Matter, 2012. **8**(17): p. 4584–4594.
14. Kirtil, E. and M.H. Oztop, *1 H nuclear magnetic resonance relaxometry and magnetic resonance imaging and applications in food science and processing*. Food Engineering Reviews, 2016. **8**(1): pp. 1–22.

2 Food Microstructure and Quality Changes in Foods during Processing

2.1 INTRODUCTION

The structure and properties of food materials are complex and heterogeneous. A fundamental understanding of the microstructure is required to predict and describe food quality changes during drying. Food security is a significant concern in many countries of the world. It is reported that one-third of the global food production, or 1.3 billion tons, is lost annually due to the lack of proper processing [1,2]. Food waste means not only the loss of food and essential nutrition but also the waste of the valuable resources used in the production of food such as land, water, energy and labour. If the produced food is not consumed, it leads to unnecessary CO_2 emissions that are a significant contributor to today's global warming problem [3]. Therefore, proper food processing must be emphasised to help reduce this massive loss, promote food security, reduce global warming and combat hunger [4].

Processing of food is a preservation method that inhibits the growth of bacteria, yeasts and mould through the removal of water/moisture content [5]. Processed foods have gained commercial importance, and their growth on an industrial scale has become an essential sector of the agricultural industry [6,7]. Heat and mass transfer in food tissue during drying depends on the mechanical properties at different levels of structure, that is the cellular level (i.e. the architecture of the tissue cells and their interaction) and the organ level (i.e. the arrangement of cells into tissues and their chemical and physical interactions). Therefore, a proper understanding of cellular-level water distribution in the raw food material and its evolution during processing is crucial to understand and accurately describe dehydration processes [8,9]. Therefore, the primary aim of this chapter is to discuss the cellular structure of food tissue and the changes in the cellular structure during the drying process.

2.2 THE MICROSTRUCTURE OF PLANT-BASED FOOD MATERIALS

The structure of the food materials includes protein, glucose, water, air and minerals. The food microstructure is a complex combination of these components [10,11]. Cell

DOI: 10.1201/9781003047018-2

walls are formed by fibres, cellulose, pectin and hemicellulose. The combination of the cells, cell wall and the air space build the food tissue. The air space inside the tissue, commonly known as the intercellular space, contributes to the porosity of the food materials [12]. The cells have high water permeability, and the food tissue retains water by the surface tension effect. The membrane inside the cell wall is known as plasmalemma. This membrane is semi-permeable and responsible for the osmotic effect. The cells contain cytoplasm, tonoplast and vacuole, which are the major contributors to the cells' metabolic activities. The vacuole contains most of the water of the cells and creates hydrostatic pressure. This hydrostatic pressure ensures the firm cellular structure of the food materials. This pressure is known as the turgor pressure of the cells. It also has a strong relationship with the rheological properties of the food tissue. The loss of cellular water from the vacuole directly impacts the microstructure of the food materials. The structural features of the food material at the cellular level and the tissue level are different, and the difference is explained in Figure 2.1. The structural features at the macro-level and the micro-level are different. A hierarchical structure of a food material is presented in Figure 2.1. The microstructure of food materials can be defined as the spatial arrangement of the cells and pores. The structure of the food material comprises three basic components: polysaccharides, proteins and lipids. The cellular materials are mainly composed of glucose.

2.3 THE RELATIONSHIP BETWEEN FOOD STRUCTURE AND FOOD PROPERTY

Drying conditions have a significant influence on the microstructural changes of the food being processed. The modification of microstructures can be regulated by controlling drying conditions. Novel approaches to the drying process mostly rely on the proper understanding of the microstructures' architecture and organisation.

Food drying is a complex system, and in order to properly understand the process, an in-depth knowledge of the food structure and composition is required. It is also important to understand the evolution of the food materials for designing an efficient food processing system. As a result of the poor understanding of the structure–property relationship of food materials, most of the time the food engineers end up with designing inefficient processes. Sometimes the empirical data and experience help the designer to design the process, but it is a costly and often inaccurate effort. Therefore, food processing becomes more costly and less efficient. The lack of knowledge of the structure–property relationship also leads to the poor quality of the processed food.

The cellular structure of the food materials plays a vital role in the transport properties and stability during processing. The major structural properties that play a vital role in the quality of the food materials are porosity, shrinkage and diffusivity. Shrinkage of the food materials is a physical phenomenon that occurs during food processing due to the removal of the water from the cells. It has a major influence on the quality of the food product as it reduces the wettability and adsorption capability and modifies the texture. The microstructure of the food material is deformed when shrinkage occurs. Sometimes the microstructure

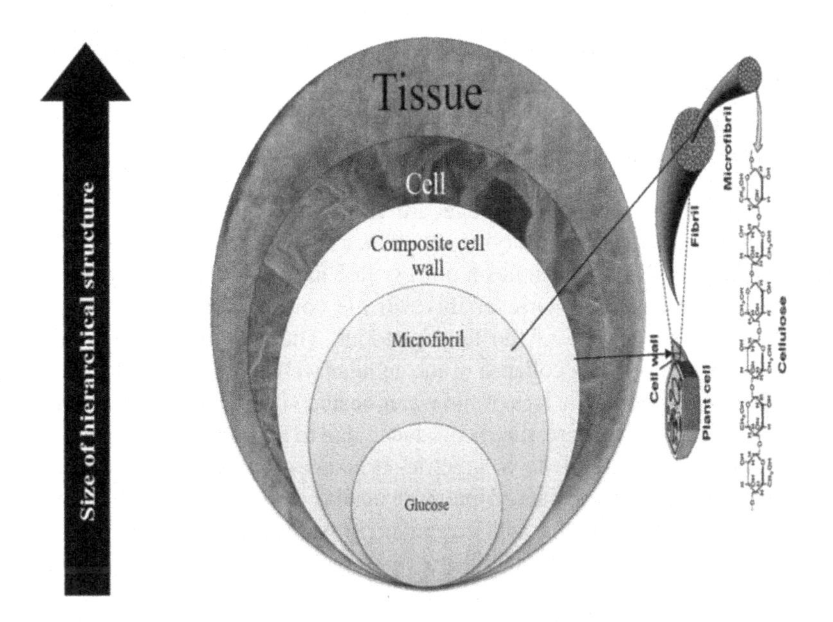

FIGURE 2.1 Structural features of the food materials at the cellular and tissue levels [13].

may collapse during shrinkage. The deformation and the collapse are dependent on the processing conditions [14,15].

The most important rheological properties of the food materials for governing the shrinkage phenomenon are the elastic modulus (modulus of elasticity), hardness and stiffness. Elastic modulus is the measure of the resistance of food materials to elastic deformation when stress is applied to them. On the other hand, hardness is a measure of the resistance to plastic deformation, penetration, indentation and scratching [16]. These variations of the modulus of elasticity are mainly due to the variations of the moisture content and the structural heterogeneities of food materials at different stages of food processing. Large deformation of anisotropic shrinkage occurs sometimes during food processing. This happens due to the variation of the elastic properties and moisture loss. This large shrinkage, which is initiated at the cellular level, plays a vital role in the food quality [17]. To address this phenomenon, it is required to develop an accurate food processing model. However, currently, not all cellular properties to develop an accurate model are available in the literature. Therefore, an extensive experimental investigation is needed to facilitate the development of an accurate food process model.

Porosity impacts the physical properties and the quality of the processed food. Mathematically, the porosity of the food material is expressed as the ratio of the volume fraction of the intercellular space to the cellular volume. Depending on the characteristics, three types of pore can be formed: closed pore, open pore and blind pore [18]. The porosity in the food materials varies with the processing condition and the moisture content of the food.

2.4 STRUCTURAL CHANGE OF FOOD DURING DRYING

Microscopic collapse and shrinkage are the consequence of microstructural modification over the time of drying. There is a fine difference between collapse and shrinkage: collapse illustrates a process in which the cellular- or tissue-level structure may break down irreversibly; on the contrary, shrinkage refers to a shortening in the volume of the food material. The structural change of food materials is the result of a collapse of the cellular wall at the cellular level during the drying period when the loss of the water occurs significantly [19]. During the drying process, negative pressure is produced due to the removal of water, which is the major reason behind the volumetric shrinkage. A pore allows air inside the food for maintaining a balanced pressure in the environment. But due to the collapse of the pore of the cell, negative pressure is derived, and shrinkage occurs [20]. Both shrinkage and collapse resist producing pores during the drying process, but the shrinkage can be minimised by producing a sufficient amount of pores, which can improve the rehydration property effectively. However, owing to the migration of the ongoing moisture, the collapse of the cells and intercellular spaces is increased [21].

During the migration of water, the plant-based tissue shows a tendency to compensate for the void developed due to the migrated water by shrinkage and collapse. However, a fraction of the volume remains intact as porosity after the migration of water. At the early stage of the drying process, it is challenging to predict the accurate degree of porosity although some researchers assumed a volumetric shrinkage, which is nearly equal to the removed water from the food sample [22]. Nevertheless, this smooth and linear relationship between volumetric shrinkage and moisture reduction has not always been available in the literature [23].

There are several important factors upon which the nature of collapse and shrinkage is dependent, such as the porosity of the tissue, intercellular adhesion and the strength of cell walls. Tissues having strong intercellular adhesion and lower porosity may fail by collapse, whereas the opposite phenomena occur in the case of tissues with high porosity [24]. At the beginning of the drying stages, shrinkage occurs without collapsing, but in the final stages, collapse occurs. In summary when mass reduction from the food material is equal to the volume reduction of the solid matrix, it can be defined as the ideal shrinkage. It may be only possible if no mechanical restriction hinders the process.

On the other hand, different types of intermittent drying essentially cause different internal moisture transfer mechanism and heat transfer mode [25]. The internal moisture transfer mechanisms directly or indirectly affect the various forms of microstructure changes in the food product. However, it is very difficult to generalise any relationship between the changes in microstructure and particular conditions of drying. Such a notion is also true for the intermittency of any secondary energy sources. In the following section, the change of microstructure throughout intermittent microwave-convective drying has been presented to illustrate how intermittency affects the microstructure of food materials. Temperature and moisture distribution evolution play active roles in modifying the food microstructure. In general, high temperature and moisture regions show less porosity and more rupture. On the other hand, low temperature and moisture regions depict uniform porous microstructure. Temperature distribution on the surface of the samples obtained from the thermal imaging camera for convective drying at 70°C, continuous microwave drying (CMD) and intermittent microwave convective drying (IMCD) is presented in Figure 2.2.

The figure shows that in the case of convective drying, higher temperature exists at the edge of the sample and gradually decreases towards the centre. Over the drying time, temperature distribution shows the same trend in hot air drying of the sample. Unlike hot air drying, a random temperature distribution has been observed during microwave drying. This non-uniform distribution is the result of the uneven electromagnetic distribution. In addition to this, a higher temperature exists at the core of the sample.

Like microwave heating, IMCD causes a higher temperature in the interior of the sample. However, the temperature is redistributed and has a more uniform pattern in IMCD than in CMD. Therefore, intermittent application of microwave reduces the impact of non-uniform temperature distribution, resulting in a better overall quality of food materials [26]. How this temperature distribution affects

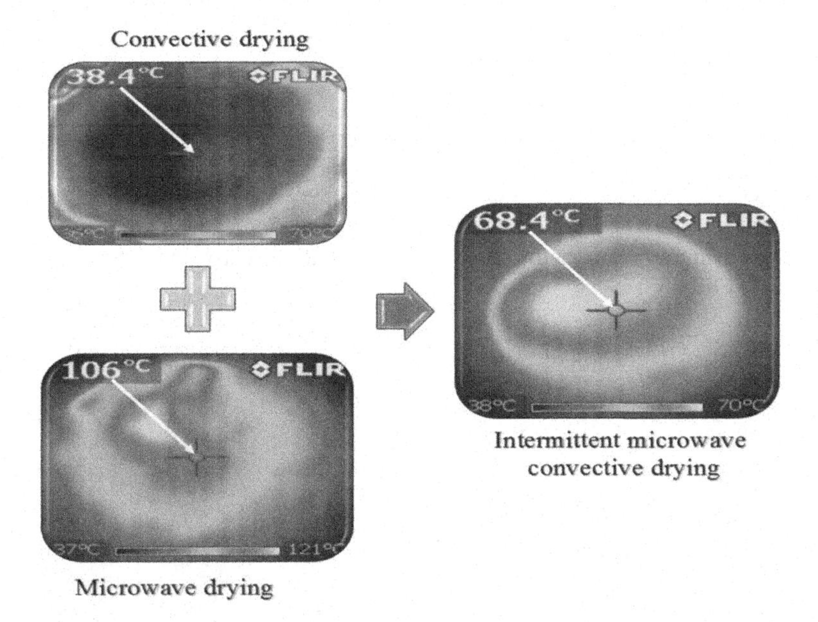

FIGURE 2.2 Temperature distribution in an apple slice during three selected drying processes. The top-left temperature manifests the temperature of the pointed area, and the colour legend is presented at the bottom of each image [7].

the structural modification of the sample during drying is discussed in the following sections. Generally, in hot air drying of fruits and vegetables, tissues are featured by extensive shrinkage and microstructural changes, such as fruit 'softening' or loss of 'firmness' [27].

A significant amount of cell collapse has been observed due to the thermal exposure of the cells on the surface of the sample. With the optimum balance of the volumetric heating and energy flow to the surface, the casehardening can be avoided. The volumetric heating (microwave heating) may cause puffing within the food sample, and an increase of vapour pressure leads to the porosity of the plant-based food materials. Microwave drying causes overheating, which results in the burning of solid materials. When the food materials get exposed to microwaves, resulting in local overheating, the warmer areas are better able to absorb further energy than the colder areas. It may be due to the lower heat exchange rate between the hot spots and the rest of the material, moisture content and the mobility of the dipolar component. Similarly, due to the non-uniform higher temperature inside the sample, a rapid phase change occurs. Compared to convective drying, IMCD provides a better product quality, as shown in Figure 2.3. Since the last stage of drying is the least efficient part of the conventional drying system, the incorporation of microwave energy would be optimised if it is used to remove the last one-third of the moisture content [28]. It can be claimed that the use of microwave energy in the early stages of drying may result in cellular collapse and bulk shrinkage in the final products [29]. However, it is still not conclusive when the microwave energy should be applied to get the best finished product. Other studies [30] recommended that microwave application at the finishing stage leads to a more porous structure.

2.5 EFFECT OF MICROSTRUCTURE CHANGES ON QUALITY

This section describes how product quality changes with the change of microstructure, considering colour as one of the quality indicators. Food colour is considered an important factor for appetite stimulation. Through colour, a person

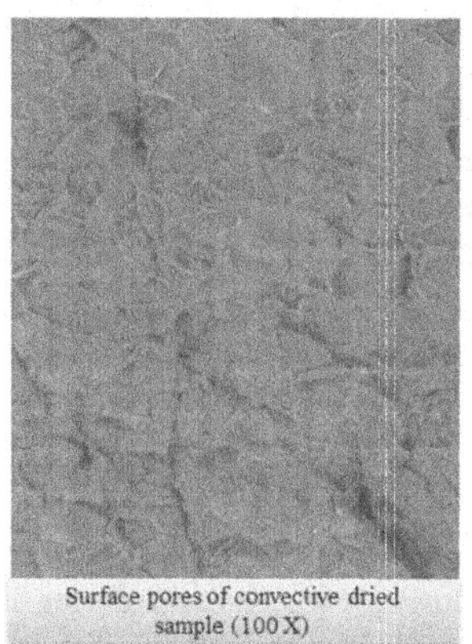

Surface pores of convective dried sample (100X)

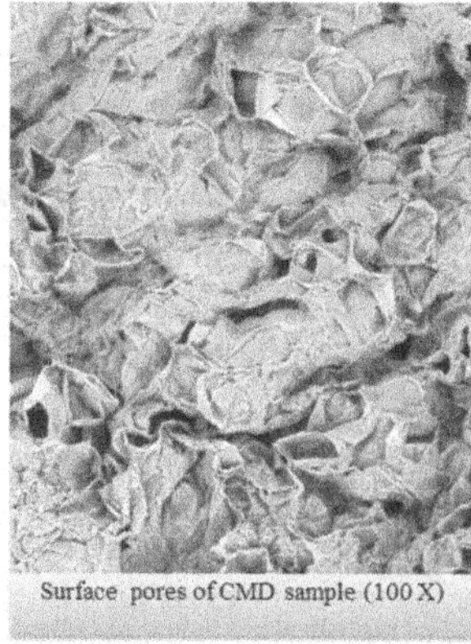

Surface pores of CMD sample (100X)

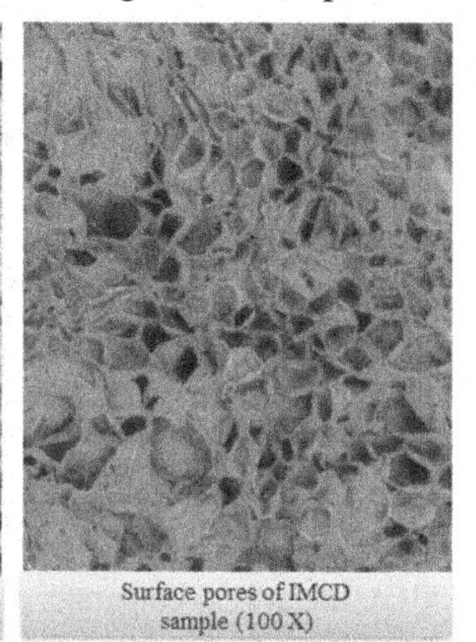

Surface pores of IMCD sample (100X)

FIGURE 2.3 Morphology of dried apple after convective, CMD and IMCD drying.

can have an idea or indication about the flavour of the food. When looking at the fruit, its colour helps as an indicator of its quality; for example, a pink strawberry colour indicates that the taste is not too strong, a green strawberry indicates the bitterness of the fruit, a black or brown colour indicates that the strawberry is deteriorated, while a nice red colour states the high quality of the strawberry. For these reasons, colour change during food drying, manufacturing and storing has become a common interest in industries. During drying, colour can be considerably affected by pigment, enzymatic activity or Maillard nonenzymatic browning reaction. The changes in porosity and surface texture affect the reflectance of light on the food surface. When the light strikes the surface of an object, it is either transmitted, absorbed or reflected. The reflected light indicates the colour. Improper application of drying conditions results in a significant change in the roughness of the surface of the processed food as the result of the modification of the microstructure. Hence, the original gloss of food will be reduced, which decreases the brightness of the product, making the product darker and less attractive. Various types of fruits and vegetables differ in structure and colour. In our everyday fruit consumption, we consume a massive amount of pigments such as carotenoids, anthocyanins and chlorophylls. These pigments have different chemical and physical properties and are affected by various factors such as light, pH change, fruit processing, contamination, oxidation, time and processing temperature.

The food's appearance, colour, taste and texture are being analysed, and many experiments have been conducted to maintain high food quality during and after food processing. During the processing, the structure of the dried fruit changes and the colour is modified, and its size is reduced. Increased exposure to hot air or high temperature leads to pigment destruction and non-enzymatic browning reactions; hence, the quality of food is reduced. The change in the microstructure and colour during the convective drying of persimmon is shown in Figure 2.4. It was reported that, as drying progressed, the colour of the sample turned from reddish-orange to dark brown in persimmon. In the process, the sample moisture content is reduced; thereby, the rate at which the moisture migrates from the inner to outer surface got decreased the evaporation rate at the sample surface. This led to the increase in the surface temperature, accelerating the browning reaction rate. Moreover, the dryness of the surface increased with the drying time, leading to the more visible roughness of the sample. As a result wider pores became more evident, which also affected the reflectance of light on the sample.

2.6 CONCLUSIONS

The food microstructure evolution and property changes with the drying process have been presented in this chapter. It is clear that the microstructural features of the food materials mainly govern the changes in the properties of the food materials. This chapter also explained the different physical food properties including porosity and shrinkage, as well as their effects on the quality of food during processing. Generic modelling of food processing is complex due to the

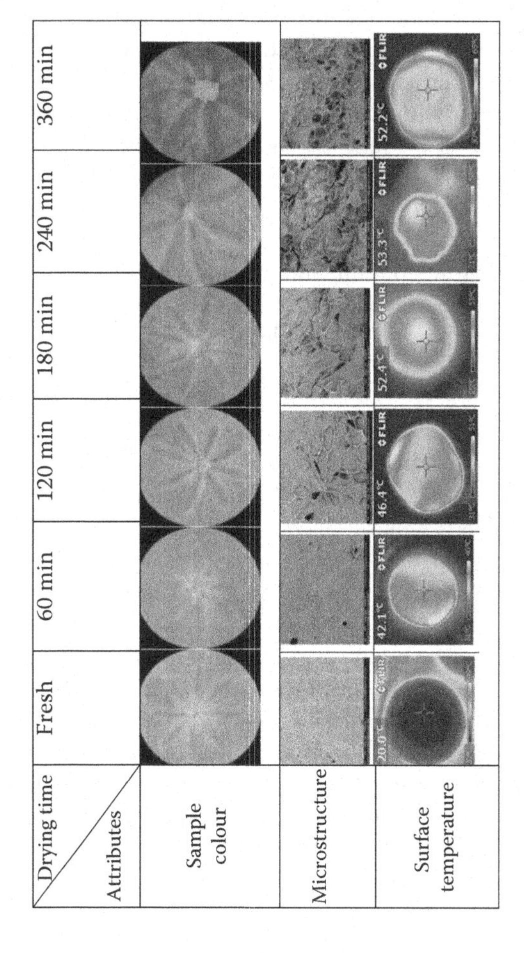

FIGURE 2.4 Persimmon microstructure and colour evolution during processing (drying).

complex structure–property relationship of each food materials. The processing conditions also play a vital role in the evolution of the properties. Therefore, a detailed information regarding the properties and their evolution during processing is required to develop an accurate food processing model. The utmost importance for the food industries is to retain the fundamental functionality and maintain the quality of the food materials during processing. Therefore, extensive experimental investigation at the cellular level needs to be done to facilitate information to the food engineers.

REFERENCES

1. Abesinghe, A., et al., *Effects of ultrasound on the fermentation profile of fermented milk products incorporated with lactic acid bacteria*. International Dairy Journal, 2019. **90**: pp. 1–14.
2. Argyropoulos, D., A. Heindl, and J. Müller, *Assessment of convection, hot-air combined with microwave-vacuum and freeze-drying methods for mushrooms with regard to product quality*. International Journal of Food Science & Technology, 2011. **46**(2): pp. 333–342.
3. Bolin, H.R. and C.C. Huxsoll, *Scanning electron microscope/ image analyzer determination of dimensional postharvest changes in fruit cells*. Journal of Food Science, 1987. **52**(6): pp. 1649–1650.
4. Devahastin, S. and C. Niamnuy, *Modelling quality changes of fruits and vegetables during drying: a review*. International Journal of Food Science and Technology, 2010. **45**(9): pp. 1755–1767.
5. Dhalsamant, K., P.P. Tripathy, and S.L. Shrivastava, *Effect of pretreatment on rehydration, colour and nanoindentation properties of potato cylinders dried using a mixed-mode solar dryer*. Journal of the Science of Food and Agriculture, 2017. **97**(10): pp. 3312–3322.
6. Duc Pham, N., et al., *Quality of plant-based food materials and its prediction during intermittent drying*. Critical Reviews in Food Science and Nutrition, 2019. **59**(8): pp. 1197–1211.
7. Fanta, S.W., et al., *Microscale modeling of coupled water transport and mechanical deformation of fruit tissue during dehydration*. Journal of Food Engineering, 2014. **124**: pp. 86–96.
8. Joardder, M.U., et al., *Factors affecting porosity*. In Porosity. 2016, Springer. pp. 25–46.
9. Joardder, M.U.H., C. Kumar, and M.A. Karim, *Prediction of porosity of food materials during drying: Current challenges and directions*. Critical Reviews in Food Science and Nutrition, 2018. **58**(17): pp. 2896–2907.
10. Joardder, M.U.H., C. Kumar, and M.A. Karim, *Food structure: Its formation and relationships with other properties*. Critical Reviews in Food Science and Nutrition 2015. In Press.
11. Joardder, M.U., et al., *A micro-level investigation of the solid displacement method for porosity determination of dried food*. Journal of Food Engineering, 2015. **166**: pp. 156–164.
12. Karim, M.A. and M. Hawlader, *Drying characteristics of banana: theoretical modelling and experimental validation*. Journal of Food Engineering, 2005. **70**(1): pp. 35–45.

13. Kumar, C., et al., *A porous media transport model for apple drying*. Biosystems Engineering, 2018. **176**: pp. 12–25.
14. Khan, M.I.H., C., Kumar, M.U.H., Joardder, and M.A. Karim, *Multiphase porous media modelling: A novel approach of predicting food processing performance*. Critical Reviews in Food Science and Nutrition, 2016. https://doi.org/10.1080/10408398.2016.1197881
15. Karim, M.A. and M.N.A. Hawlader, *Mathematical modelling and experimental investigation of tropical fruits drying*. International Journal of Heat and Mass Transfer, 2005. **48**(23–24): pp. 4914–4925.
16. Kumar, C., et al., *Investigation of intermittent microwave convective drying (IMCD) of food materials by a coupled 3D electromagnetics and multiphase model*. Drying Technology, 2018. **36**(6): pp. 736–750.
17. Khan, M.I.H. and M. Karim, *Cellular water distribution, transport, and its investigation methods for plant-based food material*. Food Research International, 2017. **99**: pp. 1–14.
18. Khan, M.I.H., S.A. Nagy, and M. Karim, *Transport of cellular water during drying: An understanding of cell rupturing mechanism in apple tissue*. Food Research International, 2018. **105**: pp. 772–781.
19. Khan, M.I.H., et al., *Fundamental understanding of cellular water transport process in bio-food material during drying*. Scientific Reports, 2018. **8**(1): pp. 15191.
20. Kumar, C. and M. Karim, *Microwave-convective drying of food materials: A critical review*. Critical Reviews in Food Science and Nutrition, 2019. **59**(3): pp. 379–394.
21. Kumar, C., et al., *Temperature redistribution modelling during intermittent microwave convective heating*. Procedia Engineering, 2014. **90**(0): pp. 544–549.
22. Lozano, J.E., E. Rotstein, and M.J. Urbicain, *Shrinkage, porosity and bulk density of foodstuffs at changing moisture contents*. Journal of Food Science, 1983. **48**: pp. 1497–1502, 1553.
23. Mahiuddin, M., et al., *Shrinkage of food materials during drying: Current status and challenges*. Comprehensive Reviews in Food Science and Food Safety, 2018. **17**(5): pp. 1113–1126.
24. Ormerod, A.P., et al., *The influence of tissue porosity on the material properties of model plant tissues*. Journal of Materials Science, 2004. **39**(2): pp. 529–538.
25. Pakowski, Z. and R. Adamski, *Formation of underpressure in an apple cylinder during convective drying*. Drying Technology, 2012. **30**(11-12): pp. 1238–1246.
26. Rahman, M., et al., *A micro-level transport model for plant-based food materials during drying*. Chemical Engineering Science, 2018. **187**: pp. 1–15.
27. Rahman, M.M., M.U. Joardder, and A. Karim, *Non-destructive investigation of cellular level moisture distribution and morphological changes during drying of a plant-based food material*. Biosystems Engineering, 2018. **169**: pp. 126–138.
28. Rahman, M.M., et al., *Multi-scale model of food drying: Current status and challenges*. Critical Reviews in Food Science and Nutrition, 2018. **58**(5): pp. 858–876.
29. Wang, N. and J.G. Brennan, *Changes in structure, density and porosity of potato during dehydration*. Journal of Food Engineering, 1995. **24**(1): pp. 61–76.
30. Zhang, M., et al., *Trends in microwave-related drying of fruits and vegetables*. Trends in Food Science & Technology, 2006. **17**(10): pp. 524–534.

3 Use of Advanced Micro-imaging Methods in Food Processing

3.1 INTRODUCTION

Most food materials have a heterogeneous microstructure, and understanding the microstructure is very important to explain the micro-level transport phenomena and morphological changes during food processing [1,2] as these micro-level changes significantly impact the product quality [3,4]. This chapter presents a brief overview of the micro-imaging technologies used to determine the micro-level morphological changes and transport during food processing [5,6]. The transport phenomena in food material during food processing are mainly initiated from the cell level. All the changes that are observed at the macro level start at the cell level [7,8]. Cells have unique properties, and these properties need to be considered for the investigation of food processing.

A comprehensive mathematical model or extensive experimentation can interpret the micro-level food processing [9–11]. However, appropriate micro-level experiments are still required to develop an accurate mathematical model [12]. Investigating the cellular-level properties and selecting and designing an appropriate experimental method are significant challenges [13]. Some studies are found in the literature to investigate the cellular-level properties and water distribution during food processing [7,13].

The major focus of this chapter will be on the advancement of the cellular-level experimental investigation process in the field of food processing technology. Some popular methods available for the micro-level experimental investigation are X-ray micro-tomography [14,15], light microscopy [16], synchrotron X-ray imaging [17], nuclear magnetic resonance (NMR) [18], magnetic resonance imaging (MRI) [19], scanning electron microscopy (SEM) [20], atomic force microscopy (AFM) [21] and near-infrared reflectance (NIR) spectroscopy [22]. However, each method has its own advantages and disadvantages, and a particular method may not be very useful for investigating some properties or phenomena due to the limitations and the special features of that method [11]. NMR and X-ray CT can be useful for investigating spatial and temporal moisture distribution [23]. AFM is useful for nanoindentation experiments for determining mechanical properties. For obtaining, real-time cellular-level information during food processing, electrostatic sensor technology can be beneficial. It is important to be extra cautious before selecting any

DOI: 10.1201/9781003047018-3

suitable technique considering its shortcomings. It is important to be careful before selecting a method or technique as every method has its own limitations.

3.2 EXPERIMENTAL TECHNIQUES FOR THE MICRO-LEVEL IMAGING OF FOOD PROCESS

For investigating the cellular-level properties, several methods are available. In the following sections, the most suitable experimental methods will be discussed briefly as each method will be discussed in detail in the later chapters of this book.

3.2.1 Atomic Force Microscopy

Atomic force microscopy is a very popular method for investigating the mechanical and rheological properties of the food material. It works based on the interaction between the small tip at the end of a cantilever and the sample surface (see Figure 3.1). The deflections of the cantilever can be measured by the optical reflecting laser beam. The deflection can be horizontal or vertical. The AFM can perform three major investigations including topographic imaging, manipulation and force measurement. It can take high-resolution three-dimensional images up to the nanoscale for food materials. The AFM detector measures the deflection of the cantilever by converting it into an electrical signal.

The researchers are attracted to use the AFM method for investigating the morphological changes of plant and animal tissues. The rheological properties of the living cells and the extracellular matrix were also investigated using AFM [24]. To provide comprehensive information on food processing by evaluating the spoilage and related structural modification, AFM has been used as a very effective tool [25]. The stiffness of the cells [26] and the formation of the solid bridge in spray-dried sodium carbonate particles during the spray drying process were also studied by the AFM [27]. The characterisation of cells and tissues of the food materials can be done by the AFM method [24]. Furthermore, AFM can measure the adhesive binding force of the rotting enzyme and the spoilage microorganism of food materials [28]. The study of the adhesive capacity of the microbes helps increase the shelf life of the processed food. This kind of study is very important in the food processing industry. The effectiveness of various preservation methods can also be measured by the AFM. Even though the AFM has many benefits, it has not been widely used yet in the food processing industry due to the slow scanning speed and a limited magnification range.

3.2.2 X-ray Micro-tomography

X-ray tomography offers an alternative visualisation of the food microstructure as it is non-invasive and non-destructive and no sample preparation is required [11,15,29]. It can evaluate the microstructural changes up to the nanometres level. X-ray computed tomography (CT) was introduced in the clinical area in

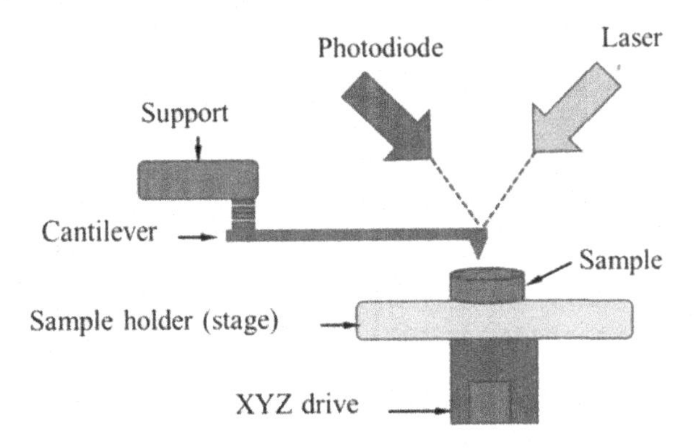

FIGURE 3.1 AFM working principle.

1970, while it was introduced in food processing in 1991. A schematic diagram of the acquisition principle of X-ray tomography is presented in Figure 3.2. The transverse conical beam from the object is recorded in the detector. The magnification of the sample is dependent on the position of the sample between the source and the detector. Numerous photographs are usually recorded for accurate data acquisition from X-ray images.

Cellular-level water distribution and the morphological change during the processing of plant-based food materials were investigated by X-ray tomography technique [15,31]. The effect of the processing condition on the microstructure of the food materials was invested in that study [15]. The posterity and the variation of the porosity of food materials during processing were also investigated by the X-ray μCT imaging method [32].

The tomographic images contain information on the internal structure and composition of the food materials. Three-dimensional sample volume can be created from the tomographic information by the different types of reconstruction algorithm. Special software packages are required for the data analysis and the reconstruction purpose. The density and composition variation of the sample can also be derived from the tomographic images.

3.2.3 Nuclear Magnetic Resonance

NMR spectroscopy is a robust method, and it can analyse the food sample at the molecular level rapidly. The NMR experimentation process does not require purification or separation steps, making this method very popular in the food industry. NMR spectroscopy has been developed based on the magnetic properties of certain nuclei. The nuclei have an even mass number with an odd atomic number or an odd mass number. These nuclei have a magnetic moment that does not have a macroscopic analogue, leading to a sequential spin, which makes the nuclei electrically charged [7].

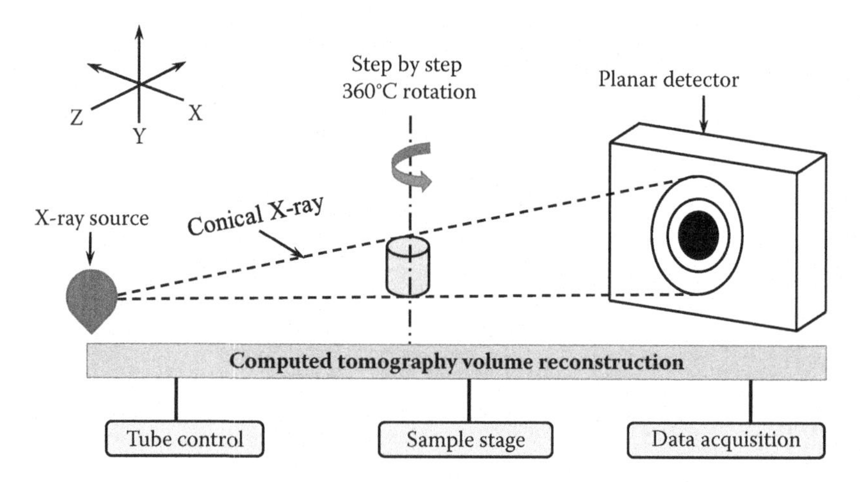

FIGURE 3.2 Simple schematic of the mechanism of X-ray microcomputed tomography [30].

In the field of food processing, 1D and 2D frequency dimensions are commonly used. The reflected NMR signals from the food sample change based on the magnetic field. The NMR images can be filtered with the selection of the image sequence timing. The timing sequence is dependent on the types of food tissue. The NMR imaging process is graphically illustrated in Figure 3.3.

NMR spectroscopy is one of the most robust experimental methods in the food process industry due to its capacity for evaluating and analysing the solid and liquid components. For the measurement of the cellular-level water distribution in the plant-based food material, this method was extensively used. The water distribution in the food materials during processing and microstructural heterogeneity of the food materials can also be measured by the NMR–T_2 relaxometry. Determining the water distribution in the food materials with the NMR spectroscopy requires low-resolution NMR. Therefore, it was very popular in the food drying industry. The use of the high-resolution NMR technique has been increased recently. With the high resolution, NMR can analyse the solid and the liquid state matrices and the compositions of the materials as well. This capability leads to the higher use of NMR in food processing industries, particularly in quality control, structural characterisation and sensory evolution. For this reason, the food processing industries can now meet the standards of the regulatory body and the consumers by providing high-quality food materials.

NIR spectroscopy is a non-destructive analytical technique and one of the most prominent technologies for online product analysis in food processing industries. This technology operates based on the operation of electromagnetic radiation and the wavelength of the radiation between 700 and 2,500 nm. The food processing industries use the NIR spectra for quality control as well as process control. For example, NIR technology was used to investigate the water content and protein content in meat processing industries. This technology was

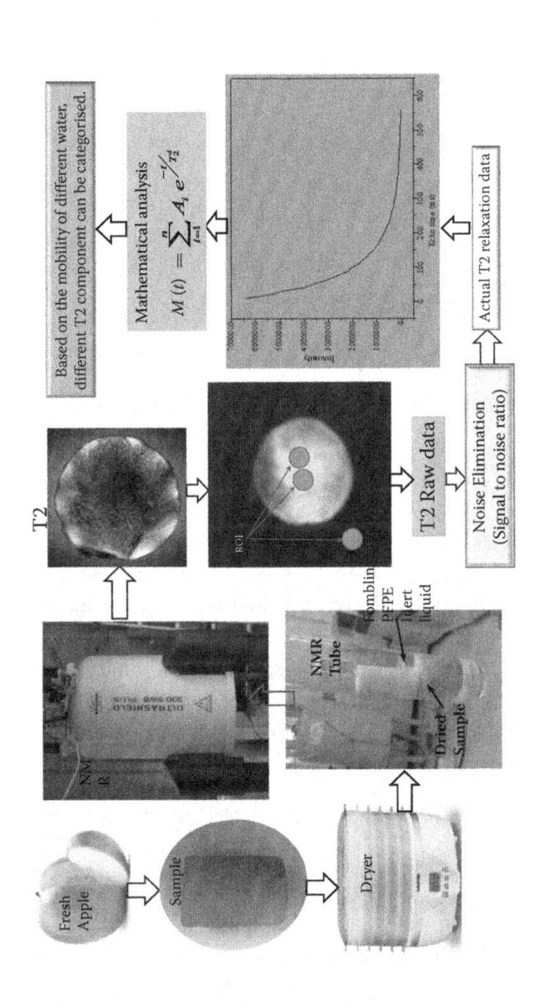

FIGURE 3.3 Schematic diagram of the NMR technique for the determination of water distribution in food tissue [13].

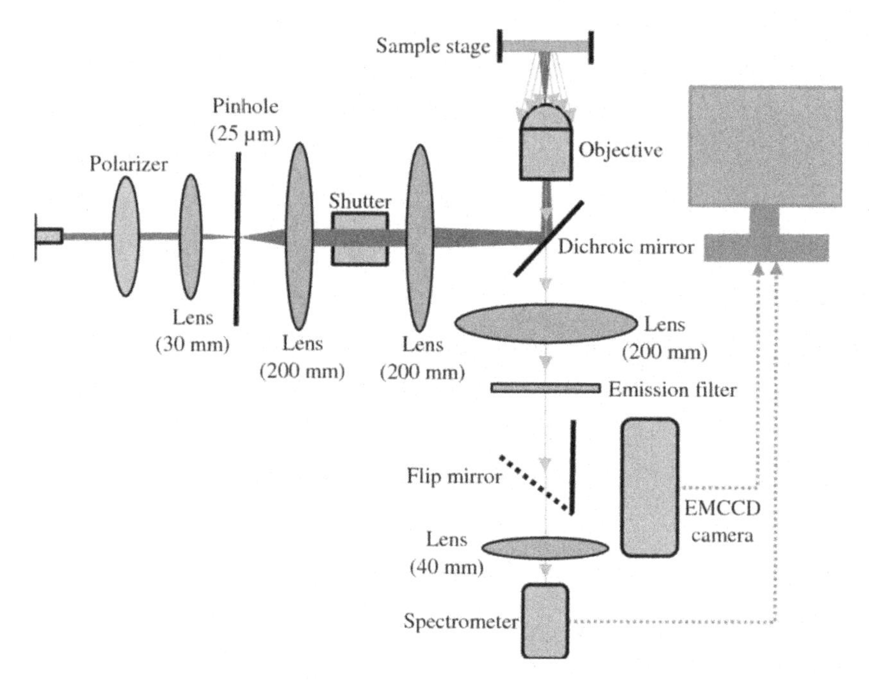

FIGURE 3.4 Schematic diagram of NIR spectroscopy [33].

also used to analyse the moisture content in the grain processing industries. The major advantage of the NIR technology is that it does not require any sample preparation. The major components of the NIR are lens, beam slitters, reflector, CCD camera and the detector, as shown in Figure 3.4.

NIR technology is a very useful tool for online investigation and analysis of food material; however, it has some limitations. The major limitation of this technology is that it requires a comparatively large sample, which restricts the use of this technology in cellular-level analysis.

3.2.4 Light Microscopy

Light microscopy is also a popular tool for investigating the cellular morphology of food materials. This technology has a big potential in the application of the food processing industries. The use of light microscopy started in the nineteenth century by plant cell researchers. However, since the twentieth century this technology has been widely used by food researchers. The major benefit of light microscopy is cost-effectiveness. The main part of a light microscope includes lighting equipment, lens, supporting stages and sample stages. Nowadays compound light microscopes are also available, which included double lenses. The double lenses facilitate more magnification of the food samples. In earlier stages, light microscopy has been used in food science for determining the contamination or adulteration of foods. Later, the microstructure of the food materials and the structure–property relationship were investigated by the light microscopy technique. The quality of the food materials is very important during processing, and light microscopy can be a vital tool to meet the quality requirements in the industry.

3.3 CONCLUSION

Micro-imaging technologies are essential development for investigating the micro-level properties of the food materials in food processing industries. This chapter presented a brief review regarding the modern micro-imaging techniques that can be very useful for the food processing industries. Among these methods, not all of them were widely used in food processing industries, yet these have a great potential. Depending on the requirements and the facilities, appropriate technology should be chosen. It is also important to choose the appropriate technology while considering the limitations and the benefit of each technology.

REFERENCES

1. Kumar, C., et al., *Investigation of intermittent microwave convective drying (IMCD) of food materials by a coupled 3D electromagnetics and multiphase model*. Drying Technology, 2018. **36**(6): pp. 736–750.
2. Kumar, C., et al., *A porous media transport model for apple drying*. Biosystems Engineering, 2018. **176**: pp. 12–25.
3. Abesinghe, A., et al., *Effects of ultrasound on the fermentation profile of fermented milk products incorporated with lactic acid bacteria*. International Dairy Journal, 2019. **90**: pp. 1–14.
4. Duc Pham, N., et al., *Quality of plant-based food materials and its prediction during intermittent drying*. Critical Reviews in Food Science and Nutrition, 2019. **59**(8): pp. 1197–1211.
5. Joardder, M.U., C. Kumar, and M. Karim, *Prediction of porosity of food materials during drying: Current challenges and directions*. Critical Reviews in Food Science and Nutrition, 2018. **58**(17): pp. 2896–2907.
6. Joardder, M.U. and M. Karim, *Development of a porosity prediction model based on shrinkage velocity and glass transition temperature*. Drying Technology, 2019. **37**(15): pp. 1988–2004.
7. Khan, M.I.H., et al., *Cellular level water distribution and its investigation techniques*. In: Chung-Lim L., Azharul K. (eds). Intermittent and Nonstationary Drying Technologies: Principles and Applications, 2017, CRC Press. pp. 193–210.
8. Khan, M.I.H., C., Kumar, M.U.H., Joardder, and M.A. Karim, *Multiphase porous media modelling: A novel approach of predicting food processing performance*. Critical Reviews in Food Science and Nutrition, 2016. https://doi.org/10.1080/10408398.2016.1197881
9. Rahman, M.M., et al., *A micro-level transport model for plant-based food materials during drying*. Chemical Engineering Science, 2018. **187**: pp. 1–15.
10. Khan, M.I.H., et al., *Fundamental understanding of cellular water transport process in bio-food material during drying*. Scientific Reports, 2018. **8**(1): pp. 1–12.

11. Rahman, M.M., et al., *Multi-scale model of food drying: Current status and challenges.* Critical Reviews in Food Science and Nutrition, 2018. **58**(5): pp. 858–876.
12. Joardder, M.U., et al., *A micro-level investigation of the solid displacement method for porosity determination of dried food.* Journal of Food Engineering, 2015. **166**: pp. 156–164.
13. Khan, M.I.H. and M. Karim, *Cellular water distribution, transport, and its investigation methods for plant-based food material.* Food Research International, 2017. **99**: pp. 1–14.
14. Welsh, Z., et al., *Multiscale modeling for food drying: State of the art.* Comprehensive Reviews in Food Science and Food Safety, 2018. **17**(5): pp. 1293–1308.
15. Rahman, M.M., M.U.H. Joardder, and A. Karim, *Non-destructive investigation of cellular level moisture distribution and morphological changes during drying of a plant-based food material.* Biosystems Engineering, 2018. **169**: pp. 126–138.
16. Autio, K. and M. Salmenkallio-Marttila, *Light microscopic investigations of cereal grains, doughs and breads.* LWT-Food Science and Technology, 2001. **34**(1): pp. 18–22.
17. du Roscoat, S.R., et al., *Estimation of microstructural properties from synchrotron X-ray microtomography and determination of the REV in paper materials.* Acta Materialia, 2007. **55**(8): pp. 2841–2850.
18. Khan, M.I.H., et al., *Investigation of bound and free water in plant-based food material using NMR T2 relaxometry.* Innovative Food Science & Emerging Technologies, 2016. **38**: pp. 252–261.
19. McCarthy, M.J., *Magnetic resonance imaging in foods.* 2012, Springer Science & Business Media.
20. Kuo, J., *Electron microscopy: Methods and protocols.* Vol. 369. 2007, Springer Science & Business Media.
21. Kirby, A.R., et al., *Visualization of plant cell walls by atomic force microscopy.* Biophysical Journal, 1996. **70**(3): pp. 1138–1143.
22. Nicolai, B.M., et al., *Nondestructive measurement of fruit and vegetable quality by means of NIR spectroscopy: A review.* Postharvest Biology and Technology, 2007. **46**(2): pp. 99–118.
23. Mahiuddin, M., et al., *Shrinkage of food materials during drying: Current status and challenges.* Comprehensive Reviews in Food Science and Food Safety, 2018. **17**(5): pp. 1113–1126.
24. Thomas, G., et al., *Measuring the mechanical properties of living cells using atomic force microscopy.* Journal of Visualized Experiments, 2013(76): p. e50497. https://doi.org/10.3791/50497
25. Zhao, L., N. Kristi, and Z. Ye, *Atomic force microscopy in food preservation research: New insights to overcome spoilage issues.* Food Research International, 2020: p. 110043. https://doi.org/10.1016/j.foodres.2020.110043
26. Solon, J., et al., *Fibroblast adaptation and stiffness matching to soft elastic substrates.* Biophysical Journal, 2007. **93**(12): pp. 4453–4461.
27. Leaper, M.C., et al., *Solid bridge formation between spray-dried sodium carbonate particles.* Drying Technology, 2012. **30**(9): pp. 1008–1013.
28. Liu, Q. and H. Yang, *Application of atomic force microscopy in food microorganisms.* Trends in Food Science & Technology, 2019. **87**: pp. 73–83.
29. Mujumdar, A.S. and W. Zhonghua, *Thermal drying technologies—Cost-effective innovation aided by mathematical modeling approach.* Drying Technology, 2007. **26**(1): pp. 145–153.

30. Khan, M.I.H., M. Rahman, and M. Karim, *Recent advances in micro-level experimental investigation in food drying technology*. Drying Technology, 2020. **38**(5–6): pp. 557–576.
31. Khan, M.I.H., S.A. Nagy, and M. Karim, *Transport of cellular water during drying: An understanding of cell rupturing mechanism in apple tissue*. Food Research International, 2018. **105**: pp. 772–781.
32. Wang, Y., et al., *Pore network drying model for particle aggregates: Assessment by X-ray microtomography*. Drying Technology, 2012. **30**(15): pp. 1800–1809.
33. Karimi, N., *Direct laser writing of fluorescent microstructures containing silver nanoclusters in polyvinyl alcohol films*. 2015. https://doi.org/10.13140/RG.2.1.3080.5524

4 Use of Atomic Force Microscopy (AFM) in Food Processing

4.1 INTRODUCTION

The food microstructure is mainly composed of assemblies of irregular cellular compartments, and the dynamic variation of the mechanical and the rheological properties during the processing of food material play a vital role in the quality attributes [1,2]. Researchers have long been trying to determine the changes in the mechanical and rheological properties of food during processing [3,4]. The atomic force microscope (AFM) was invented to work as a scanning near-field tool for nanoscale investigations. It is a useful tool for investigating the cellular-level mechanical properties of different food materials. In AFM, a sharp tip is used instead of a light or electron beam. AFM can detect changes at a spatial resolution up to the sub-nanometre level as the curvature's tip radius is on the nanometre scale. During the tests, an indenter is derived into the material surface, and, subsequently, the images of the impression are taken. This technique is known as nanoindentation, which can be applied to investigate properties such as material hardness, Young's modulus, strain-rate sensitivity and material stiffness. In this technique, elastic deformation occurs on the surface of the sample because of the identification of the material. Depending on the sample types and the applied load's magnitude, the deformation can be observed as an elastic or plastic range. Plastic deformation appears while the load is increased above the elastic limit, which results in the non-linear characteristic of the loading curve [5,6]. During the unloading process, only the elastic portion of the displacement is recovered, and the irreversible plastic deformation finally makes the surface of the material form an indentation.

The structure of foods is complex and highly structured, comprising cells, intercellular spaces, the cell membrane and the cell wall [7,8]. The water inside food materials is distributed in different proportions in different parts of the food microstructure [9,10]. The major component of the cell pectin and two cells are separated by the middle lamella [11]. The microstructure and the composition element play a vital role in the stability and mechanical properties of the food materials [6]. Especially, the cell wall has a major impact on the stability of the food microstructure [12]. The major component of the cell walls is lignin, pectin, hemicellulose and cellulose. The mechanical properties and the texture of the food materials including stiffness, hardness, modulus of elasticity and rigidity are dependent on the component [13].

DOI: 10.1201/9781003047018-4

The processing of food significantly changes the properties as the food materials face various processing conditions and moisture loss [14]. Sometimes the food materials change their state from a soft state to a hard state [15]. The change of the state of the food materials is dependent on the processing condition, and it affects the quality of the processed food [16,17]. The rheological properties of the food materials are altered due to the state change [18]. The major output of this state change is the anisotropic shrinkage of the food materials [19].

For analysing the rheological and mechanical properties of the food materials, the nanoindentation method has been used [20]. The rheological and mechanical properties of the food materials including hardness and stiffness can be obtained from the nanoindentation load–displacement output curve. The viscoelastic properties of Rosette leaves were investigated by the nanoindentation method [18]. Even the viscoelastic properties of the polymer materials were investigated by this method. For determining the structure–property relationship, structural heterogeneity characterisation is important. The structural characterisation is also possible by the AFM. Furthermore, the rheological and mechanical properties of wood cells [21], animal liver [22] and polyurethane coating film [23] can be investigated by the AFM. The AFM nanoindentation has recently been used in the food processing sector. Studies on the onion cell wall and the impact of the composition element (i.e. pectin) have been conducted by the researchers [24]. Nanomechanical properties of the apple tissue were investigated by the AFM nanoindentation method [25].

The main idea of the nanoindentation method involves the indentation of the cell's surface with an AFM tip known as indenters. The shape of the indenters can be square, rectangular and triangular [26]. The properties can be calculated from the load–displacement output curve of the AFM nanoindentation method. The details of this method are presented in the following section.

4.2 THEORY AND OPERATION OF THE AFM NANOINDENTATION

The major component of atomic force microscopy (AFM) is a micro-cantilever with a probe, a detector, piezoelectric ceramic, a motion detection device and a feedback regulator. A computer processing system is required to process the data obtained from the AFM. An AFM experimental setup picture is presented in Figure 4.1. The food sample is placed on a magnetic plate of the indenter to conduct the nanoindentation experiments. The tip of the indenter needs to be calibrated before every experiment. To ensure high-quality images, the optical camera needs to be set up with suitable magnification.

Indenter tip, commonly made of stiff material, is an important tool to get an accurate result [26]. For the indentation of a small cellular area, the indenter tip is required to be small. On the other hand, larger tips can be used for a larger area like tissue-level samples. The indentation locations are mainly dependent on the cell dimension.

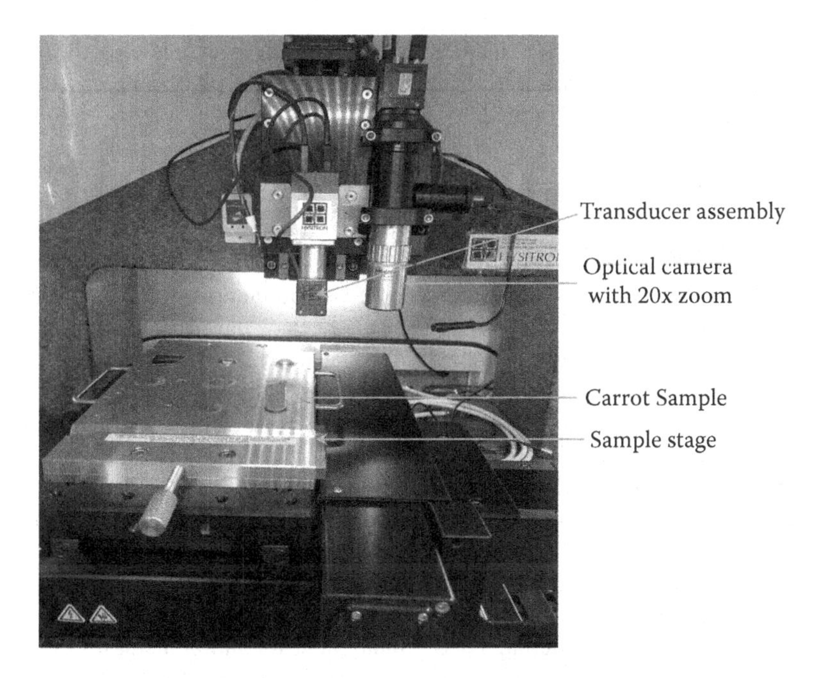

FIGURE 4.1 Nanoindentation experimental setup [27].

The load–displacement graph obtained from the indentation process needs to be analysed to get the hardness, reduced modulus and stiffness data. Young's modulus can be calculated from the hardness modulus by the following relationship [28],

$$E_r = \frac{1}{\beta}\frac{\sqrt{\pi}}{2}\frac{S}{\sqrt{A_i h_c}} \tag{4.1}$$

$$\frac{1}{E_r} = \frac{(1-\nu_i^2)}{E_i} + \frac{(1-\nu_s^2)}{E_s} \tag{4.2}$$

where S is the sample stiffness; E_r is the reduced Young's modulus; β is the geometrical constant; ν_s and E_s are the Poisson's ratio and Young's modulus, respectively; A_i is the projected indentation area $h_{c,}$ which can be calculated from the fitting polynomial of the loading–unloading curve; and ν_i and E_i are the Poisson's ratio and Young's modulus, respectively.

The cells of the food materials show viscoelastic characteristics and adhesive behaviours in the nanoindentation experiment. These characteristics of food materials lead to potential errors while determining the sample modulus. The displacement of the sample surface increases while loading and unloading are performed without any pause. To solve this issue, it is recommended to allow a

5–10 seconds pause between loading and unloading, which allows the food material to reach equilibrium. At different stages of food processing, numerous indentations should be performed at various locations of the sample.

4.3 APPLICATION OF AFM IN FOOD PROCESSING

4.3.1 Assessing Mechanical Properties during Processing

Hardness, elastic modulus and stiffness are very important rheological properties of the food material. Hardness is the measurement of the resistance to the indentation and plastic deformation of the food materials, and elastic modulus is the measurement of the resistance to elastic deformation. During food processing, the sequential plots of the mechanical properties are performed to observe simultaneous changes [29]. For example, the elastic modulus and hardness variation of a carrot cell along with the processing (convective drying) time are presented in Figure 4.2. From the figure, it can be seen that the elastic modulus increases very slowly with processing time in the earlier stages of the process, and it increases exponentially in the later stages (Figure 4.2a). The main reason for the variation is the change of the stress–strain relationship due to the moisture removal from the food material. This cellular water migration increases the cell wall tension, which ultimately increases Young's modulus [30].

The hardness variation of the carrot cell is shown in Figure 4.2b. The elastic modulus, the increment of the carrot cell's hardness, is slow in the earlier stages of the processing, and the increment is exponential at the later stages of processing. The main reason for these characteristics is the structural heterogeneity of the food materials [31]. The stiffness of the carrot cells also shows similar kinds of characteristics during the process (Figure 4.2c). The slow increment of the stiffness and hardness at the initial stage of the processing occurs due to the flexible and soft characteristics of the food materials [32,33]. The results obtained from this experiment completely agree with the literature [34].

4.3.2 Water Content and Mechanical Properties Relationship

During food processing, large deformation occurs due to changes in the mechanical properties and water loss. The large deformation of food materials initiates at the micro-level [35]. While developing the processing model, the researchers considered the food materials as non-linear elastic materials [36]. These models require mechanical properties and water content relationship as input variables, which can be obtained by the AFM method. The experimental relationship between the water content and the mechanical properties is presented in Figures 4.3 to 4.5.

The effect of the water loss on the elastic modulus of the food materials (carrot cell) is shown in Figure 4.3. This trend may be explained by the concept that, at the early stages of the processing, the viscous properties are dominant, while the elastic properties are dominant at the later stages of the processing. The

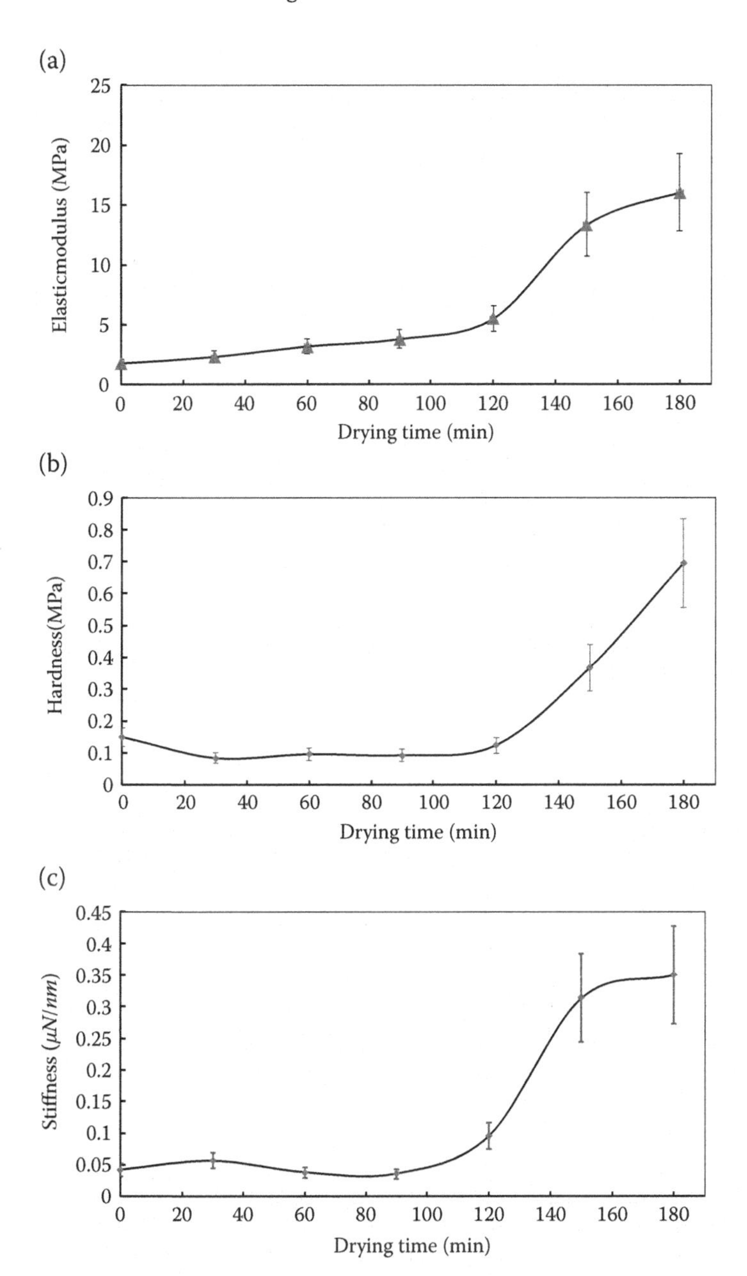

FIGURE 4.2 The variation of the mechanical properties food materials with processing (convective drying) time: (a) elastic modulus variation, (b) hardness variation and (c) stiffness variation [27].

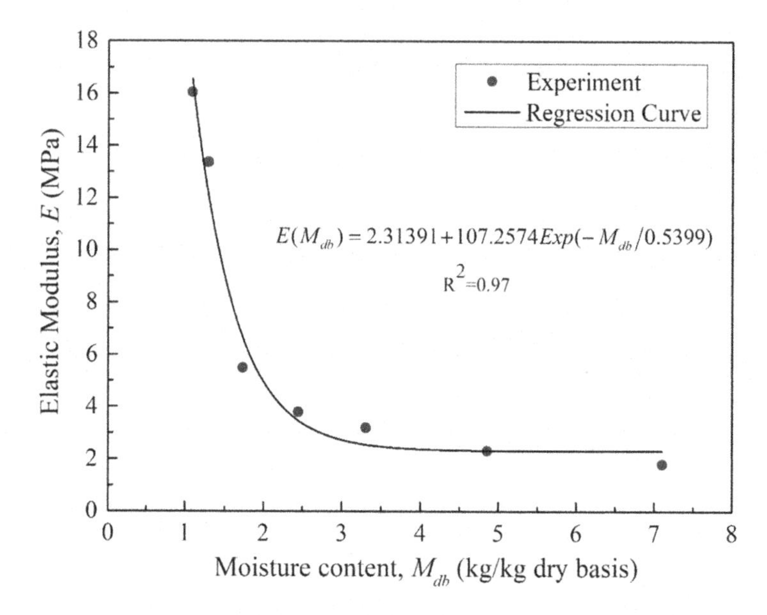

FIGURE 4.3 Water content and the elastic modulus of the relationship of the cell during food processing [27].

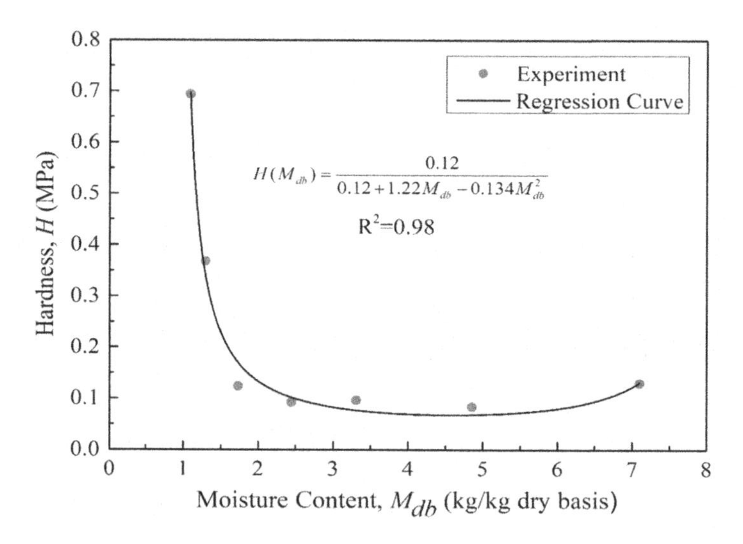

FIGURE 4.4 Relationship between water content and hardness variation of the carrot cell during processing (convective drying) [27].

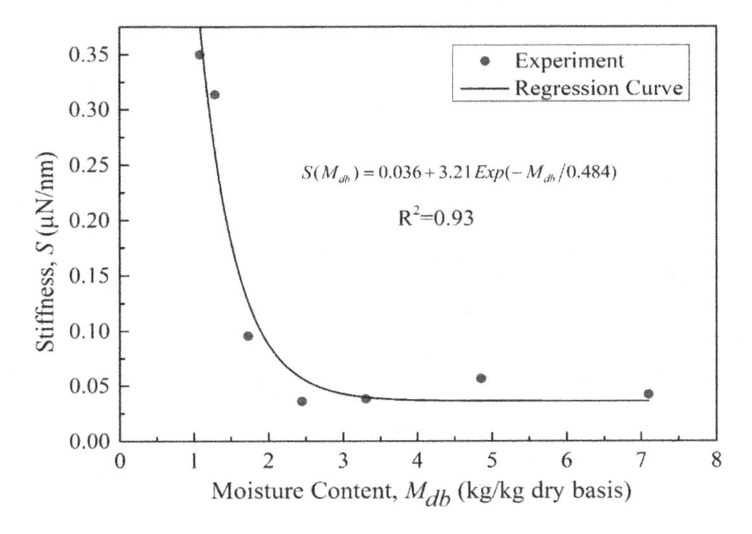

FIGURE 4.5 Water content and cell wall stiffness relationship during processing (convective drying) [27].

relationship of the water content and the elastic properties can be mathematically expressed as follows:

$$E = 2.31391 + 107.2574\exp(-1.8522M_{db}) \tag{4.3}$$

where E is the elastic modulus and M_{db} is the water content.

The good agreement of the relationship has been proven by the regression analysis value of $R^2 = 0.97$. The water content–hardness and the water content–stiffness relationship of the food material are presented in Figures 4.4 and 4.5, respectively. The mathematical expression of these relationships can be expressed as follows:

$$H = \frac{0.12}{0.12 + 1.22\,M_{db} - 0.134M_{db}^2} \tag{4.4}$$

$$S = 0.036 + 3.21\exp(-2.066M_{db}) \tag{4.5}$$

Here, S represents the stiffness of food materials and H represents the hardness of food materials.

The viscoelastic phase transition of the food materials during food processing plays a vital role in the variations of the mechanical properties [37].

4.4 CONCLUSION

The AFM technology has been proved to be a very effective method for the morphological and rheological studies of food materials. The technique is prevalent in investigating the complex food component and the variation of their mechanical properties during processing. It is possible to correlate the structure–property relationship during food processing with this powerful tool. Regression analysis can be a robust mathematical approach in this regard. The AFM experimental study reported in the literature revealed that the process environment significantly impacts the mechanical and rheological properties at the different stages of food processing. This AFM investigation can also determine the trends for cell wall stiffness and hardness of food materials during processing. Furthermore, with the regression analysis of different mechanical properties with moisture content, a set of empirical relationships can be developed. The connection will be beneficial for developing the multi-scale deformation model for food processing.

REFERENCES

1. Duc Pham, N., et al., *Quality of plant-based food materials and its prediction during intermittent drying*. Critical Reviews in Food Science and Nutrition, 2019. **59**(8): pp. 1197–1211.
2. Rahman, M.M., et al., *A micro-level transport model for plant-based food materials during drying*. Chemical Engineering Science, 2018. **187**: pp. 1–15.
3. Kumar, C., et al., *A porous media transport model for apple drying*. Biosystems Engineering, 2018. **176**: pp. 12–25.
4. Mahiuddin, M., et al., *Development of fractional viscoelastic model for characterizing viscoelastic properties of food material during drying*. Food Bioscience, 2018. **23**: pp. 45–53.
5. Mahiuddin, M., et al., *Shrinkage of food materials during drying: Current status and challenges*. Comprehensive Reviews in Food Science and Food Safety, 2018. **17**(5): pp. 1113–1126.
6. Joardder, M.U., et al., *A micro-level investigation of the solid displacement method for porosity determination of dried food*. Journal of Food Engineering, 2015. **166**: pp. 156–164.
7. Khan, M.I.H., et al., *Fundamental understanding of cellular water transport process in bio-food material during drying*. Scientific Reports, 2018. **8**(1): pp. 15191.
8. Rahman, M., et al., *Multi-scale model of food drying: Current status and challenges*. Critical Reviews in Food Science and Nutrition, 2018. **58**(5): pp. 858–876.
9. Khan, M.I.H., S.A. Nagy, and M. Karim, *Transport of cellular water during drying: An understanding of cell rupturing mechanism in apple tissue*. Food Research International, 2018. **105**: pp. 772–781.
10. Rahman, M.M., M.U. Joardder, and A. Karim, *Non-destructive investigation of cellular level moisture distribution and morphological changes during drying of a plant-based food material*. Biosystems Engineering, 2018. **169**: pp. 126–138.
11. Lopez-Sanchez, P., et al., *Rheology and microstructure of carrot and tomato emulsions as a result of high-pressure homogenization conditions*. Journal of Food Science, 2011. **76**(1): pp. E130–E140.

12. Joardder, M.U., et al., *Effect of cell wall properties on porosity and shrinkage of dried apple*. International Journal of Food Properties, 2015. **18**(10): pp. 2327–2337.
13. Pascua, Y., H. Koç, and E.A. Foegeding, *Food structure: Roles of mechanical properties and oral processing in determining sensory texture of soft materials*. Current Opinion in Colloid & Interface Science, 2013. **18**(4): pp. 324–333.
14. Kumar, C., et al., *Investigation of intermittent microwave convective drying (IMCD) of food materials by a coupled 3D electromagnetics and multiphase model*. Drying Technology, 2018. **36**(6): pp. 736–750.
15. Rahman, M.S., Non-equilibrium states and glass transitions in fruits and vegetables. In Non-equilibrium States and Glass Transitions in Foods. 2017, Elsevier. pp. 241–252. https://doi.org/10.1016/B978-0-08-100309-1.00013-4
16. Rahman, M.S., *Toward prediction of porosity in foods during drying: A brief review*. Drying Technology, 2001. **19**(1): pp. 1–13.
17. Rahman, M.S., *A theoretical model to predict the formation of pores in foods during drying*. International Journal of Food Properties, 2003. **6**(1): pp. 61–72.
18. Joardder, M.U.H., C. Kumar, and M.A. Karim, *Food structure: Its formation and relationships with other properties*. Critical Reviews in Food Science and Nutrition, 2017. **57**(6): pp. 1190–1205.
19. Telis, V.R.N., J. Telis-Romero, and A.L. Gabas, *Solids rheology for dehydrated food and biological materials*. Drying Technology, 2005. **23**(4): pp. 759–780.
20. Oyen, M.L., *Nanoindentation of biological and biomimetic materials*. Experimental Techniques, 2013. **37**(1): pp. 73–87.
21. Gindl, W., et al., *Mechanical properties of spruce wood cell walls by nanoindentation*. Applied Physics A, 2004. **79**(8): pp. 2069–2073.
22. Liu, D., et al., *Effect of ligation on the viscoelastic properties of liver tissues*. Journal of Biomechanics, 2018. **76**: pp. 235–240.
23. He, J., et al. Mechanical properties improvement of waterborne polyurethane coating films after rewetting and drying. In The Proceedings of the 5th Asia-Pacific Drying Conference: (In 2 Volumes). 2007. World Scientific.
24. Xi, X., S.H. Kim, and B. Tittmann, *Atomic force microscopy based nanoindentation study of onion abaxial epidermis walls in aqueous environment*. Journal of Applied Physics, 2015. **117**(2): pp. 024703.
25. Zdunek, A., et al., *The stiffening of the cell walls observed during physiological softening of pears*. Planta, 2016. **243**(2): pp. 519–529.
26. Liu, K.-K. and K.-T. Wan, Cells and membranes. In: Michelle, L.O. (ed). Handbook of Nanoindentation. 2019, Jenny Stanford Publishing. pp. 325–349.
27. Khan, M.I.H., et al., *Characterisation of mechanical properties of food materials during drying using nanoindentation*. Journal of Food Engineering, 2021. **291**: pp. 110306.
28. Oliver, W.C. and G.M. Pharr, *An improved technique for determining hardness and elastic modulus using load and displacement sensing indentation experiments*. Journal of Materials Research, 1992. **7**(6): pp. 1564–1583.
29. Flores-Johnson, E., et al., *Microstructure and mechanical properties of hard Acrocomia mexicana fruit shell*. Scientific Reports, 2018. **8**(1): pp. 1–12.
30. Steudle, E., U. Zimmermann, and U. Lüttge, *Effect of turgor pressure and cell size on the wall elasticity of plant cells*. Plant Physiology, 1977. **59**(2): pp. 285–289.
31. Radotić, K., et al., *Atomic force microscopy stiffness tomography on living Arabidopsis thaliana cells reveals the mechanical properties of surface and deep cell-wall layers during growth*. Biophysical Journal, 2012. **103**(3): pp. 386–394.
32. Ozturk, O.K. and P.S. Takhar, *Physical and viscoelastic properties of carrots during drying*. Journal of Texture Studies, 2019. **51**(3). pp. 532–541.

33. Martynenko, A. and M.A. Janaszek, *Texture changes during drying of apple slices*. Drying Technology, 2014. **32**(5): pp. 567–577.
34. Thussu, S. and A. Datta, *Fundamentals-based quality prediction: Texture development during drying and related processes*. Procedia Food Science, 2011. **1**: pp. 1209–1215.
35. Fanta, S.W., et al., *Microscale modeling of coupled water transport and mechanical deformation of fruit tissue during dehydration*. Journal of Food Engineering, 2014. **124**: pp. 86–96.
36. Joardder, M.U., C. Kumar, and M. Karim, *Prediction of porosity of food materials during drying: Current challenges and directions*. Critical Reviews in Food Science and Nutrition, 2018. **58**(17): pp. 2896–2907.
37. Rahman, M.S., *State diagram of foods: Its potential use in food processing and product stability*. Trends in Food Science & Technology, 2006. **17**(3): pp. 129–141.

5 X-ray Micro-tomography in Food Processing

5.1 INTRODUCTION

The microstructural changes during food processing critically influence the overall quality and other food materials' characteristics [1,2]. Moisture transport through the cells, cell wall and intercellular spaces of plant-based food materials occurs at the micro-level during the processing [3]. Hence, the understanding of the micro-level transport mechanism is essential for food processing.

The structure of food materials is multi-scale, and it mostly contains water [4]. The water inside the pant-based food materials is categorised as intercellular water, cell wall water and intracellular water [5]. This classification of the water in plant-based food materials was made based on the location of the water inside the food microstructure [6]. About 85–95% of water remains in the intracellular spaces, and the rest of the water remains in the intercellular spaces and cell walls [6]. Therefore, the overall transport process is governed by the intracellular water transport process [1] The understanding of water characteristics is crucial for developing an accurate model of food processing techniques.

During food processing, water transport occurs in three pathways: apoplastic, symplastic and transcellular (Figure 5.1). Liquid water flows under the influence of capillary forces in moisture content food materials [7,8]. Intercellular water has a higher diffusivity than intracellular water in fresh food material. For this reason, the removal of the bound water requires more energy than free water. Moreover, the removal of the bound water also affects the food microstructure [9].

In the food processing industry, information regarding the food microstructure and cellular water is required to meet the quality standard. Numerous research studies have been performed at macro-level food transport, but only a few studies have been conducted at the cellular level. As the physical quality of the food materials, including porosity and shrinkage, mostly depended on the cellular-level transport mechanism, the micro-level study during food processing is very critical [9–11]. However, most of the traditional porosity determination techniques were destructive and invasive, which made these methods unsuitable for the food processing industries [11,12]. Hence, X-ray μCT appeared to be an advantageous technique to elucidate the cellular-level transport mechanism during food drying

The first applications of computed tomography were primarily in medicine. However, it rapidly became a handy tool in physics [13], biology [14], materials science [15] and multi-scale engineering [16]. Nowadays X-ray micro-tomography

DOI: 10.1201/9781003047018-5

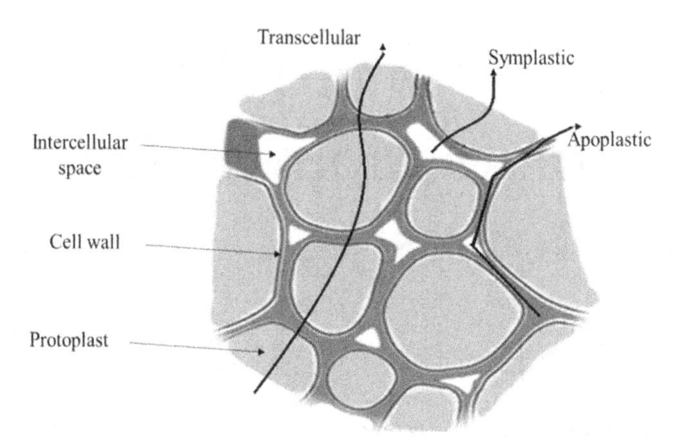

FIGURE 5.1 Transport mechanism of water inside the food microstructure [6].

technique has become a very popular tool for the characterisation of food materials in the processing industries, including apple [17], banana [18], mango [19], grains [20], kiwi [21], pears [22], carrot [23] and cucumbers [24]. Three-dimensional non-destructive reconstruction of the food materials is also possible by the X-ray μCT methods [25]. In the horticulture industries, a fresh apple tissue was characterised by the X-ray μCT [26]. Additionally, this technique was multi-scale gas transport pathway characterisation in pome fruit and apple [27]. Ripening of the fruit was also characterised to meet the quality standard by X-ray CT technique [19]. Therefore, the chapter will discuss the application of the X-ray micro-tomography technique for the experimental investigation of food materials at the cellular level during processing.

5.2 THEORY OF OPERATION

The X-ray μCT can obtain cellular-level information, such as moisture distribution, cells and pore size distribution without the chemical treatment. Therefore, the same sample can be used to analyse at the different stages of the drying process. From the technical point of view, the X-ray tomography investigation is the most appropriate method for analysing the micro-level morphological changes of food materials during processing.

Figure 5.2 shows the scanning of a sample by X-ray μCT (Scanco μCT) system with a resolution of 6 μm and a rotation of 0°–180°. A built-in software package usually controls the scanning process. The X-ray shows dual characteristics (wave particle) like the light that can be expressed as an equation of a electromagnetic wave.

$$\vartheta = \frac{c}{\lambda} \tag{5.1}$$

where c is the speed of light, ϑ is the frequency (Hz) and λ is the wavelength (m)

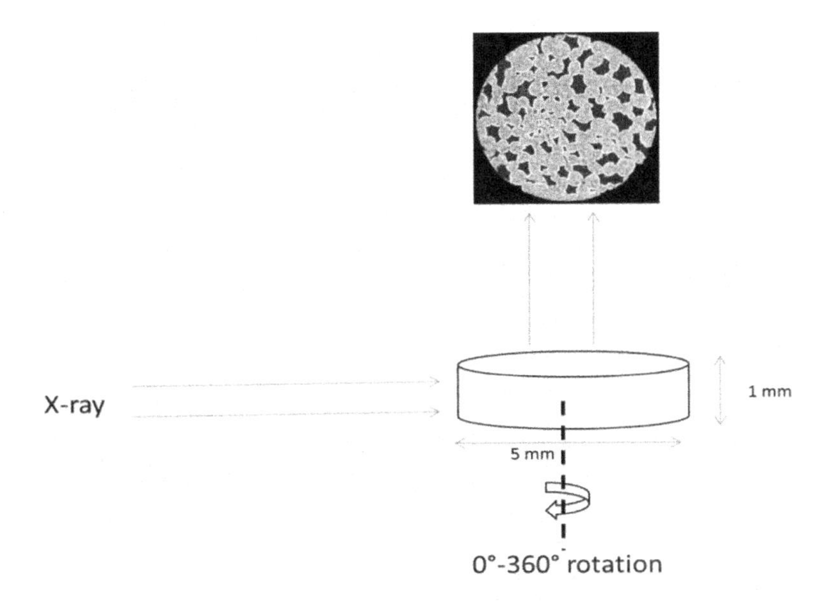

FIGURE 5.2 Schematic diagram of the X-ray μCT used for the investigation.

of the X-ray wave. The range of the X-ray wavelength used in the micro-tomography is 10^{-12} m–10^{-18} m and the corresponding frequency is 10^{16} Hz–10^{20} Hz.

An X-ray beam is directed to the sample, and the attenuated beam is recorded in a detector during the tomographic experiment. The number of a photon in the attenuated beam is directly related to the quantity of the material in the beam's path. The relationship between the transmitted and the attenuated beam is explained by the Beer–Lambert law. The mathematical expression of the Beer–Lambert law is shown in equation (5.2).

$$\mu = \rho\left(a + \frac{bZ^{3.8}}{E^{3.2}}\right) \tag{5.2}$$

where μ is the CT coefficient, ρ is the density of the material (kg/m^3), Z is the atomic number of the material, E is the energy of the transmitted beam (eKV) and a is the constant. For the reconstruction of the scanned images, a back-projection algorithm is used.

The output image obtained from the X-ray μCT is the greyscale image. The necessary information can be extracted from the images by MATLAB image processing tool. Adjustment of the greyscale image contrast is the first step of the image processing, which can be performed using MATLAB image processing tool. The image processing involves segmentation and noise. The steps involved in the image processing method are illustrated in Figure 5.3.

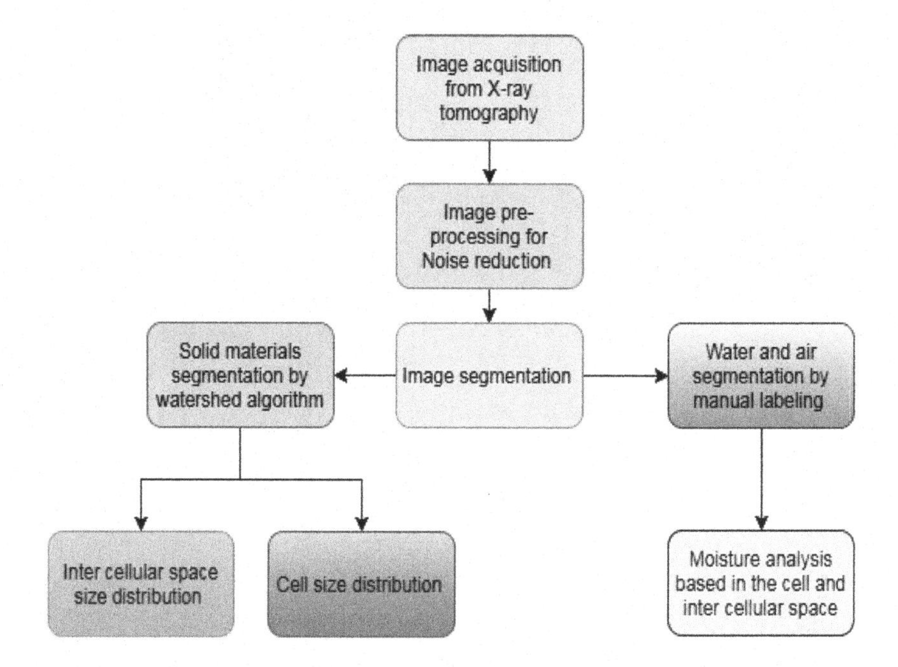

FIGURE 5.3 The steps involved in image processing in X-ray μCT.

Intensity thresholding and filtering are the major parts of the noise elimination process from the X-ray tomographic image. It is required to characterise the solid and liquid materials from the tomographic image. For this purpose, the image is segmented. For this purpose, many efficient algorithms like watershed algorithm are available in the literature [28]. From the image processing of the greyscale image, it has been found that the intensity of the liquid-filled cellular area is lower than the solid cell wall. The manual segmentation process can classify the difference between the intracellular water and the intercellular water. An example of the segmented image of the X-ray μCT is presented in Figure 5.4.

5.2.1 Water Content Measurement

In this example, image processing has been performed with the creation of the region of interest in each part of the scanned image. Each section of the images' grey level intensity is analysed to determine the μCT images' moisture content. The water content determination method from the tomographic image is available in the literature [29]. The commonly used relationship in the literature to select the moisture content is as follows:

$$\text{Moisture Content} = \frac{(\text{Grey level})_{\text{Dry solid}} - (\text{Grey level})_{\text{wet solid}}}{(\text{Grey level})_{\text{Dry solid}} - (\text{Grey level})_{\text{water}}} \quad (5.3)$$

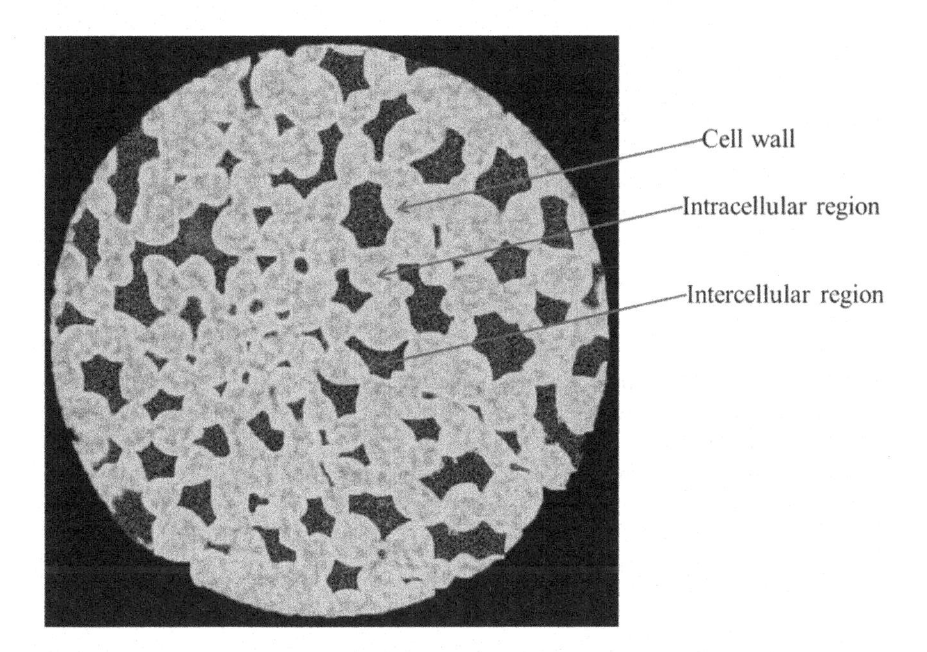

FIGURE 5.4 Segmented image of X-ray μCT.

The area of the cells and intercellular spaces are obtained from the pixel of the image. The following relationship calculated the equivalent diameter:

$$\text{Equivalent diameter} = \sqrt{\frac{4 \times area}{\pi}} \tag{5.4}$$

5.2.2 Porosity Measurement

In this example, the apples' porosity at different stages of drying was determined from the μCT images. The porosity can be calculated by the following relationship [30]:

$$Porosity = \frac{total\ volume\ of\ intercellular\ space}{Total\ volume\ of\ the\ sample} \tag{5.5}$$

5.2.3 Statistical Analysis

To find a better fit of the data and uncertainty analysis it is required to have sufficient data sets. To get a better data set, repetitive experiments at different processing conditions are recommended. Then the results can be expressed as the standard deviation and mean. The experimental results from the other tools and the results from the X-ray μ-CT images can be compared by the coefficient of determination (R^2) and standard error of estimation (SEE) analysis. This test is usually done for the accuracy of the estimated data. The coefficient of determination (R^2) is calculated using the following standard equation:

$$R^2 = \frac{[n\Sigma XY - (\Sigma X)(\Sigma Y)]^2}{[n\Sigma X^2 - (\Sigma X)^2][n\Sigma Y^2 - (Y)^2]} \tag{5.6}$$

where R^2 is the coefficient of determination, X is the moisture content determined by the moisture analyser (dry basis) and Y is the moisture content determined by the X-ray μ-CT.

The SEE was determined by using the following equation:

$$SEE = \sqrt{\frac{\Sigma(W - \bar{W})^2}{N - 2}} \tag{5.7}$$

where W is the moisture content data, $\bar{W}$ is the average value of the moisture content data and N is the number of the data.

5.3 APPLICATION OF MICRO-TOMOGRAPHY IN FOOD PROCESSING

5.3.1 Changes in the Transport Properties

The method of determining the moisture distribution during drying using X-ray μCT is explained through an example. The moisture distribution of the fresh apple slices extracted from the X-ray tomographic image is presented in Figure 5.5. The higher intensity and the lower intensity represent the cell wall, intracellular water and intercellular airspace, respectively. The intensity difference is distinguishable among the cell walls, intracellular water and intercellular airspaces. Therefore, the solid cell wall, liquid intracellular water and air-filled intercellular space can be determined by a proper image-processing algorithm. The amount of water can also be found in the tomographic image. The tomographic image clearly shows the water distribution in the fresh apple tissue.

The graphical representation of the correlation between the micro-level water distribution and the bulk moisture content during processing (convective drying at 60°C) in the apple sample is presented in Figure 5.6. In this figure, the line curve represents the bulk moisture content, and the tomographic images offer the micro-level water distribution at the different stages of processing (convective drying). The images also show the cell ruptures at various stages of drying. It is clear from the figure that the reduction of intracellular water accompanies the decrease of bulk moisture. In the earlier stage of drying, the removal of bulk moisture content is critical. In this stage,

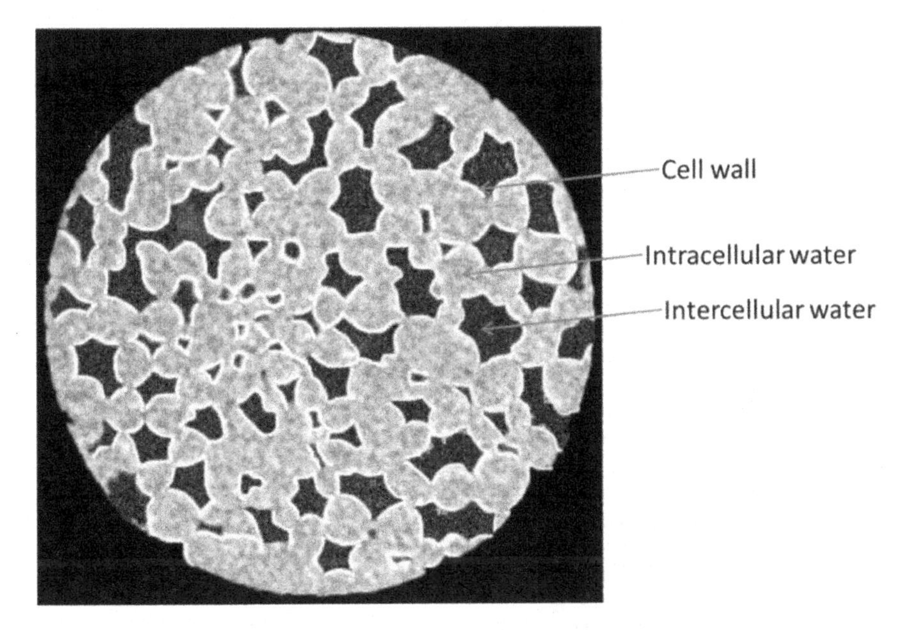

FIGURE 5.5 X-ray tomographic image of the fresh apple.

the cellular water transports through the cell wall and membrane. A considerable amount of rupture was observed after a certain amount of intracellular water removed from the cells (after 100 minutes of drying time). The long-time exposure of the food sample to the high temperature leads to the cell membrane rupture. After the cell membranes' crack, the migration of cellular water accelerates.

5.3.2 Structural Changes

The change of microstructure in food material during processing is the result of the collapse and shrinkage of the cells and cell walls due to a significant loss of water [32]. There is an acceptable difference between shrinkage and collapse. Shrinkage is a reversible reduction in the size of the food material due to processing, whereas collapse refers to the irreversible breakage of the cellular-level structure [33]. The size distribution of cells and the intercellular spaces during drying at 60°C is presented in Figure 5.7.

The size of cells and intercellular spaces was obtained from the image analysis. In the fresh sample, the cell size is bigger than the pore (Figure 5.7a). At the beginning of the drying stages, food materials shrink without cell collapse. As the drying progresses, the cell size reduces slowly while the pore size distribution changes rapidly (Figure 5.7b). During the drying process, negative pressure is created inside the cells due to the removal of water, which is the major reason behind the cellular-level deformation. The water migration process in this stage is mainly free water removal.

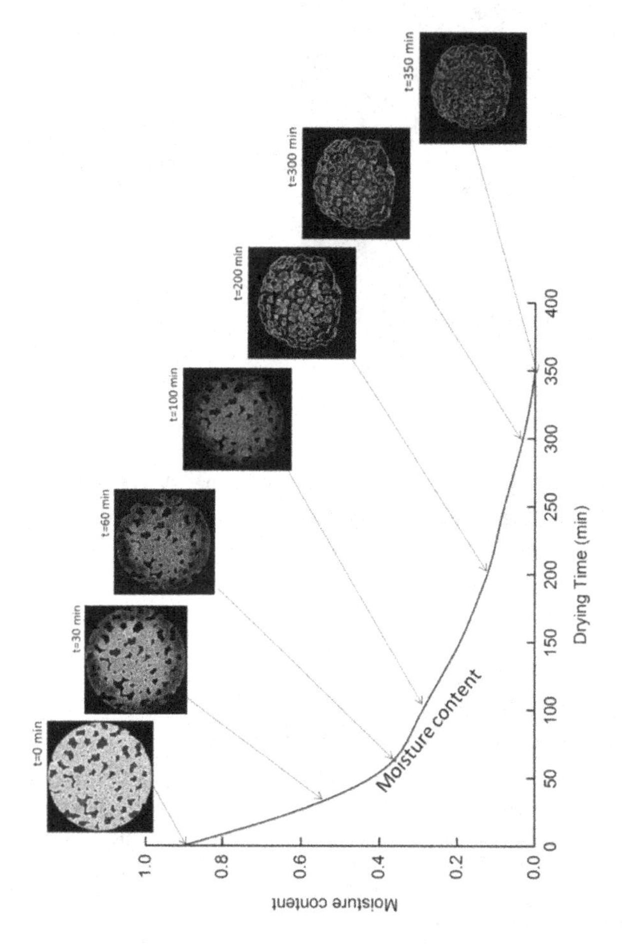

FIGURE 5.6 Cellular-level changes inside apple slices drying at 60°C [31].

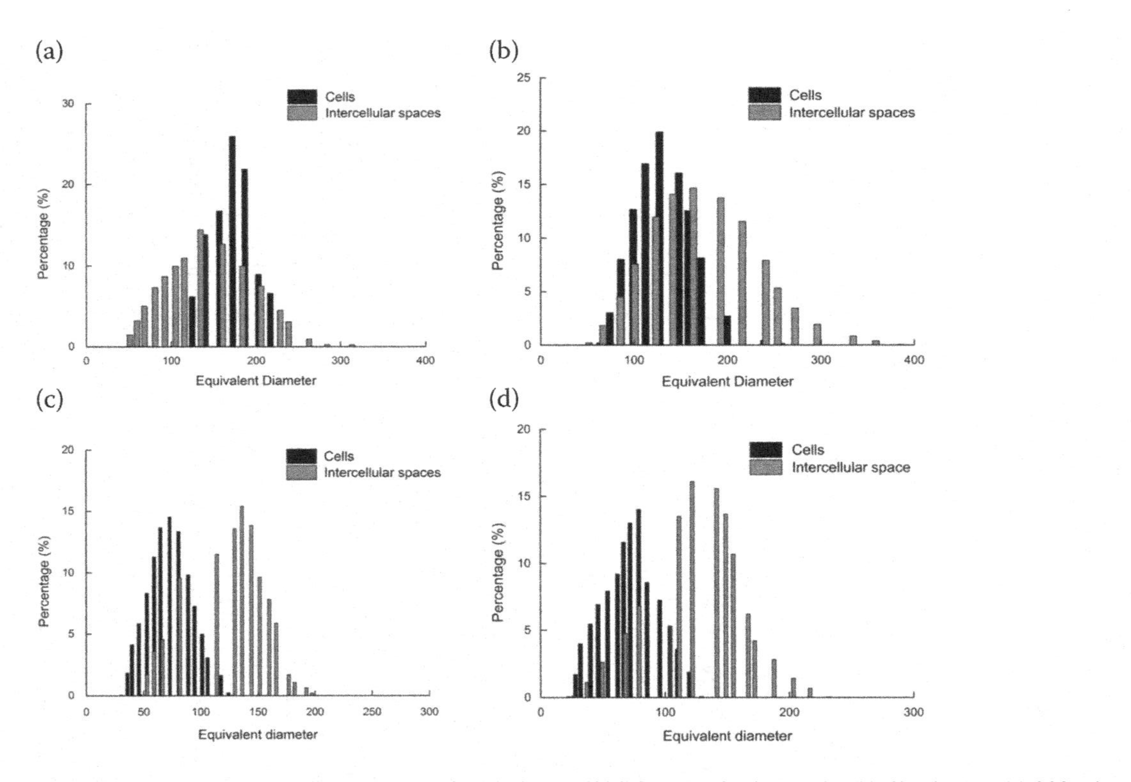

FIGURE 5.7 The size of cells and intercellular space distribution during drying at 60°C for (a) a fresh sample, (b) 60 minutes, (c) 200 minutes and (d) 350 minutes.

Owing to ongoing moisture migration, the collapse of cells and intercellular spaces is increased at the later stages of drying. After a certain drying time, cell rupture occurs progressively and decrement of the intracellular water accelerates. Therefore, it was found that the pore size distribution is greater than the cell size distributions (Figures 5.7c and d).

5.3.3 Texture Analysis

Texture analysis is considered an essential part of quality evaluation in the food processing industry. Food texture is highly related to the mechanical and rheological properties such as hardness, elasticity and viscosity. The texture is an important image feature, which can have a wide range of applications, including food quality evaluation. Changes in the intensity values of the pixels can be observed from the texture in images as the change in intensity can indicate a structural change in the food sample. Therefore, texture can be considered as an indicator of quality as it can reflect the cellular structural change in food. Textural analysis can be used in a wide range of food industries, including vegetables and fruits. X-ray μCT images are a great source for the structural analysis of food.

5.3.4 Determination of Quality

X-ray μCT has the great potential to become an essential tool in the food science and engineering field for evaluating quality and microstructure. For a better understanding of the microstructure and its changes during processing, X-ray μCT technique can be used to capture internal changes in both 2D and 3D. The use of this technique has been researched on an extensive range of food items including fish, meat, fruit and vegetables. The application of these techniques included the investigation of internal quality deterioration, mechanical and rheological characterisation, infestation detection and visualisation of pore structure and pore size distribution. Microstructural evolution in foods undergoing chemical and physical processes can be captured using X-ray.

The quality of fruit and vegetables has many dimensions, but the most widely used external indicators are colour, shape, size and surface appearance, and internal indicators are moisture content, internal bruising, decay, insect injury and colour change. One of the most important advantages of X-ray μCT imaging is that internal defects or disorders can be identified and visualised in μCT images, before they can be externally seen. Therefore, proper use of this technique should significantly reduce the storage cost, transportation cost and eventually food waste.

However, despite such immense potential, the use of this tool in the internal and external quality determination of fruit and vegetables is still relatively uncommon in the industry. Recently, 3D X-ray μCT makes its inroads in the food industry in determining the functionality of food components and internal defects in fresh and processed fruits and vegetables. X-ray μCT is an excellent method for investigating the dynamic changes in the food structure during processing. However, there are challenges in taking real- time images during processing, such as drying.

Since the first application of X-ray μCT in food science in the detection of maturity in green tomatoes [34], numerous research investigations have been performed for its applications on fruit and vegetables. The earliest applications in horticulture focussed mostly on fruits, not much on vegetables. Differences between the good- and poor-quality fruits were determined by the differences in the grey values as damaged fruit tissue appeared as darker regions and sound tissues were correlated to lighter areas. Density differences were also used to differentiate good- and bad-quality mangoes. Some researchers also used this method to study the spatial distribution of core breakdown in fruits.

Another potential application of X-ray μCT will be the non-destructive inline monitoring of fruits and vegetables, enabling the early detection of internal quality characteristics. With the 3D advantage and the ability to visualise the internal structure, an improved knowledge of products can be obtained that could result in a better understanding of the environmental effects on the fruit and vegetable structure. Therefore X-ray μCT can serve as a valuable technique for developing future prediction models for internal quality [35].

5.4 CONCLUSION

This chapter presents the application of the non-destructive X-ray μCT method in food processing using some experimental results. The analysis of cellular-level water distribution and pore size distribution and quality investigations of plant-based food material have been presented. The application has been explained by using the food drying process as an example. Food drying involves simultaneous heat, mass and momentum transfer with continuous phase change. The underlying physics of drying is closely related to the cellular properties of the material, and these cellular properties dynamically change during drying. It has been found from the example that the cell and the pore size distribution of the food material change significantly over the processing time, and this phenomenon was captured by the X-ray μCT method. This method has a huge potential to determine internal and external quality attributes of fruits and vegetables. The findings of X-ray micro-tomography investigations will help the food engineers to develop and validate an accurate transport model. It will also help the food engineers to explain the effect of the processing conditions on the food microstructure.

REFERENCES

1. Rahman, M.M., et al., *Multi-scale model of food drying: Current status and challenges*. Critical Reviews in Food Science and Nutrition, 2016. **58**(5): pp. 858–876.

2. Duc Pham, N., et al., *Quality of plant-based food materials and its prediction during intermittent drying*. Critical Reviews in Food Science and Nutrition, 2019. **59**(8): pp. 1197–1211.
3. Khan, M.I.H., et al., *Fundamental understanding of cellular water transport process in bio-food material during drying*. Scientific Reports, 2018. **8**(1): p. 15191.
4. Rahman, M.M., et al., *A micro-level transport model for plant-based food materials during drying*. Chemical Engineering Science, 2018. **187**: pp. 1–15.
5. Khan, M.I.H., et al., *Cellular level water distribution and its investigation techniques*. In Intermittent and Nonstationary Drying Technologies: Principles and Applications, 2017, CRC Press. pp. 193–210.
6. Khan, M.I.H., et al., *Investigation of bound and free water in plant-based food material using NMR T 2 relaxometry*. Innovative Food Science & Emerging Technologies, 2016. **38**: pp. 252–261.
7. Kumar, C., et al., *A porous media transport model for apple drying*. Biosystems Engineering, 2018. **176**: pp. 12–25.
8. Mahiuddin, M., et al., *Development of fractional viscoelastic model for characterizing viscoelastic properties of food material during drying*. Food Bioscience, 2018. **23**: pp. 45–53.
9. Khan, M.I.H., et al., *Experimental investigation of bound and free water transport process during drying of hygroscopic food material*. International Journal of Thermal Sciences, 2017. **117**: pp. 266–273.
10. Karunasena, H., et al., *Scanning electron microscopic study of microstructure of gala apples during hot air drying*. Drying technology, 2014. **32**(4): pp. 455–468.
11. Joardder, M.U., et al., *A micro-level investigation of the solid displacement method for porosity determination of dried food*. Journal of Food Engineering, 2015. **166**: pp. 156–164.
12. Joardder, M.U., C. Kumar, and M. Karim, *Prediction of porosity of food materials during drying: Current challenges and future directions*. Critical Reviews in Food Science and Nutrition, 2017 . https://doi.org/10.1080/10408398.2017.1345852
13. Wildenschild, D., et al., *Using X-ray computed tomography in hydrology: systems, resolutions, and limitations*. Journal of Hydrology, 2002. **267**(3): pp. 285–297.
14. Momose, A., et al., *Phase–contrast X–ray computed tomography for observing biological soft tissues*. Nature Medicine, 1996. **2**(4): pp. 473–475.
15. Salvo, L., et al., *X-ray micro-tomography an attractive characterisation technique in materials science*. Nuclear Instruments and Methods in Physics Research Section B: Beam Interactions with Materials and Atoms, 2003. **200**: pp. 273–286.
16. Wood, S.A., et al., *Measurement of three-dimensional lung tree structures by using computed tomography*. Journal of Applied Physiology, 1995. **79**(5): pp. 1687–1697.
17. Almeida, G., et al., *Physical behavior of highly deformable products during convective drying assessed by a new experimental device*. Drying Technology, 2017. **35**(8): pp. 906–917.
18. Madiouli, J., et al., *Non-contact measurement of the shrinkage and calculation of porosity during the drying of banana*. Drying Technology, 2011. **29**(12): pp. 1358–1364.
19. Cantre, D., et al., *Characterization of the 3-D microstructure of mango (Mangifera indica L. cv. Carabao) during ripening using X-ray computed microtomography*. Innovative Food Science & Emerging Technologies, 2014. **24**: pp. 28–39.
20. Zhu, L.-J., et al., *Study of kernel structure of high-amylose and wild-type rice by X-ray microtomography and SEM*. Journal of Cereal Science, 2012. **55**(1): pp. 1–5.

21. Cantre, D., et al., *Microstructural characterisation of commercial kiwifruit cultivars using X-ray micro computed tomography*. Postharvest Biology and Technology, 2014. **92**: pp. 79–86.
22. Muziri, T., et al., *Microstructure analysis and detection of mealiness in 'Forelle'pear (Pyrus communis L.) by means of X-ray computed tomography*. Postharvest Biology and Technology, 2016. **120**: pp. 145–156.
23. Voda, A., et al., *The impact of freeze-drying on microstructure and rehydration properties of carrot*. Food Research International, 2012. **49**(2): pp. 687–693.
24. Donis-Gonzalez, I.R., et al., *Internal characterisation of fresh agricultural products using traditional and ultrafast electron beam X-ray computed tomography imaging*. Biosystems Engineering, 2014. **117**: pp. 104–113.
25. Verboven, P., et al., *Three-dimensional gas exchange pathways in pome fruit characterized by synchrotron X-ray computed tomography*. Plant Physiology, 2008. **147**(2): p. 518–527.
26. Mendoza, F., et al., *Three-dimensional pore space quantification of apple tissue using X-ray computed microtomography*. Planta, 2007. **226**(3): pp. 559–570.
27. Herremans, E., et al., *Automatic analysis of the 3-D microstructure of fruit parenchyma tissue using X-ray micro-CT explains differences in aeration*. BMC Plant Biology, 2015. **15**(1): p. 264.
28. Schlüter, S., et al., *Image processing of multiphase images obtained via X-ray microtomography: A review*. Water Resources Research, 2014. **50**(4): pp. 3615–3639.
29. Léonard, A., et al., *Moisture profiles determination during convective drying using X-ray microtomography*. The Canadian Journal of Chemical Engineering, 2005. **83**(1): pp. 127–131.
30. Joardder, M.U., et al., *Porosity: Establishing the relationship between drying parameters and dried food quality*. 2015: Springer.
31. Rahman, M.M., M.U.H. Joardder, and A. Karim, *Non-destructive investigation of cellular level moisture distribution and morphological changes during drying of a plant-based food material*. Biosystems Engineering, 2018. **169**: pp. 126–138.
32. Ramos, I.N., T.R.S. Branda˜o, and C.L.M. Silva, *Structural Changes During Air Drying of Fruits and Vegetables*. Food Science Technology International, 2003. **9**(3): pp. 201–206.
33. Mahiuddin, M., et al., *Shrinkage of food materials during drying: Current status and challenges*. Comprehensive Reviews in Food Science and Food Safety, 2018. **17**(5): pp. 1113–1126.
34. Brecht, J.K., et al., *Using X-ray-computed tomography to nondestructively determine maturity of green tomatoes*. HortScience, 1991. **26**(1): pp. 45–47.
35. Schoeman, L., et al., *X-ray micro-computed tomography (μCT) for non-destructive characterisation of food microstructure*. Trends in Food Science & Technology, 2016. **47**: pp. 10–24.
36. Khan, M., et al., *Multiphase porous media modelling: A novel approach of predicting food processing performance*. Critical Reviews in Food Science and Nutrition, 2016. https://doi.org/10.1080/10408398.2016.1197881

6 Nuclear Magnetic Resonance (NMR) in Food Processing Applications

6.1 INTRODUCTION

The thermal processing of food causes many changes in the food quality [1,2]. Determining and maintaining the quality of food is of critical importance as consumer acceptability and price largely depends on the quality of fresh and processed food [3,4]. Fruits and vegetables are highly perishable as they contain 80–95% water. Therefore, a fundamental understanding of the cellular water distribution in fresh food [5] and its changes during storage, transportation and thermal processing can significantly improve our understanding of the quality changes of food [6,7]. Researchers have been working to understand the micro-level structure, water distribution and their relationship with quality [8–10]. NMR has been emerging as one of the methods to determine the cellular structure, water distribution and product quality.

NMR is a process in which nuclei with fixed magnetic moments, such as ^{15}N, ^{31}P, ^{23}Na, ^{13}C and ^{1}H, exchange energy with an alternating magnetic field [11]. In the 1950s, NMR was first used in the food science to determine the moisture content of foods [12]. NMR is now used in a variety of food fields, including numerical analysis, nutritional or functional characteristics, quality check and control of processes [13,14]. The benefits of NMR include the elimination of the need to isolate various food ingredients; minimal pre-treatment and preparation of the samples; and repeatability, non-destructiveness and quantitativeness [15]. The detection limits and sensitivity, in contrast to gas chromatography and mass spectrometry, are limitations of the current NMR technology [16].

Based on the NMR spectrum and magnetic resonance imaging (MRI) techniques, NMR technology is of two types: high-resolution NMR (HR-NMR) and low-field NMR (LF-NMR) [15]. NMR technology has been widely used in food science in recent years due to the need for effective quality control analysis and the growing need for technological and product innovation in the food industry. NMR spectroscopy is used to determine the relaxation time, and the relaxation time of NMR is used to assess the consistency of food and to measure chemical compositions [17]. NMR relaxation technology is very common in the food industry.

DOI: 10.1201/9781003047018-6

Advances in the use of NMR in foods have been investigated, including wine, dairy products, and applications in food traceability, authenticity of food, and the relationship between water delivery and quality parameters in fruits, vegetables, and meat has been studied. In recent years, much progress has been made in the study of NMR's application in the food science.

6.2 PRINCIPLE OF NMR

NMR is a powerful tool for deciding consistency parameters and tracing the cellular water environment. It is a popular method for studying atomic nuclei with an odd number of protons and/or neutrons. In a magnetic field, such nuclei have a magnetic moment that oscillates at a certain frequency known as the Larmor frequency. When located in the presence of a strong magnetic field, a small proportion of the nuclei interact with the magnetic field in an antiparallel way, similar to a magnet bar in a magnetic field. The number of spins aligning in this way is proportional to the magnetic field strength, resulting in a sample with a net magnetisation vector that can be analysed spectroscopically. Between the ground state and the aligned lower energy state, the Boltzmann relation defines the distribution of nuclei [18] according to equation (6.1):

$$\frac{N_{up}}{N_{low}} = e^{-\Delta E/kT} = e^{-hv/kT}, \tag{6.1}$$

where k is the Boltzmann constant, h is Planck's constant, the difference in energy of the two states is ΔE, v is the resonant frequency and T is the absolute temperature. For a magnetic field strength of 18.8 T, the ratio of nuclei in the two energy states is 0.999872 at thermal equilibrium and room temperature. The low sensitivity of the NMR technique is due to this limited population difference, which is exploited by the technique.

The gyromagnetic ratio of nuclei with a magnetic moment is unique, and the resonant frequency v_o is dependent on the magnetic field strength Bo [19], as shown in equation (6.2).

$$v_0 = \frac{\gamma B_0}{2\pi}, \tag{6.2}$$

The absorption of energy and coherent precession of the nuclei is caused by the application of a short radio frequency signal matched to the precession frequency of the nucleus of interest. The amplitude of the radio frequency pulse determines the degree of rotation of the magnetisation vector. The nuclei return to equilibrium when the radio frequency signal is turned off, and their spins lose coherence.

After applying the radio frequency pulse, the NMR signal is determined by putting the sample in a coil tuned to the resonance frequency of the nuclei of interest. A slight current is induced in a receiver coil by the nuclei coherently processing magnetic moment, which dissipates as the nuclei revert to equilibrium. The intensity–time signals that results can be recorded and Fourier-transformed to produce an intensity–frequency spectrum. Equation 6.3 gives the maximum magnetisation that can be detected M_o.

$$M_0 = \frac{\gamma^2 h^2 N_s B_0}{4kT}, \tag{6.3}$$

According to equation (6.3), γ is the nuclear species-specific gyromagnetic ratio, N_s is the number of spins and Bo is the magnetic field strength [18]. As can be seen from this equation, the frequency of the measured signal is proportional to the number of nuclei in the sample volume and the strength of the applied magnetic field. This behaviour produces populations of spins that correspond to the amount of electron shielding experienced by nuclei in each environment in the presence of different electronic environments. The signal dispersion due to local shielding effects is referred to as chemical change, and it is used to produce an NMR spectrum.

6.3 LIMITATIONS OF NMR APPLICATION IN BIOLOGICAL TISSUE

NMR is a unique technique for studying the various cellular compartments of water, but it has some drawbacks. The technique's poor sensitivity is one of its most significant drawbacks. To achieve a satisfactory signal-to-noise (SNR), substantial signal averaging is necessary, owing to the limited number of nuclei associated with the magnetic field. When measuring water in biological systems, the signal is reasonable, but when looking at molecules other than water, the low sensitivity becomes an issue. Measuring nuclei other than protons exacerbates the problem of low SNR due to their lower gyromagnetic ratio. High nuclei concentrations, longer acquisition times or a combination of both are needed to measure such samples.

Since the technique relies on a homogeneous magnetic field, NMR measurements are impossible for nuclei that do not have a magnetic moment, even though they are NMR sensitive. This can be disturbed at the molecular level by paramagnetic nuclei, which reduce NMR sensitivity. Powerful superconducting magnets, which are used in NMR systems, are associated with considerable costs, since constant cryogen levels are to be maintained. Despite these drawbacks, NMR experiments provide a wealth of knowledge, making the technique highly useful.

6.4 METHODS OF NMR EXPERIMENTATION

Relaxometry experiments using proton NMR (^{1}H NMR) have been useful in the study of plants and plant-based food items, in determining anatomical details of the entire tissue and, in particular, the water state. Since water protons dominate the proton signal, and the proton NMR signal amplitude is proportional to tissue proton density, ^{1}H NMR relaxometry signals, which are averaged over the entire sample, provide information on the water environments within the plant tissue [20]. The behaviour of water proton relaxation and the water exchange rates between these compartments are regulated by T_2, which is determined by the water mobility in the tissue's microscopic environment and the intensity of the applied magnetic field. The spin–spin T_2 relaxation is the transverse portion of the magnetisation vector that exponentially decays towards its equilibrium value after being excited by radiofrequency energy. It can be described by the following equation (6.4):

$$N(t) = \sum_{i=1}^{n} X_i e^{-t/T_2^i}, \tag{6.4}$$

where $N(t)$ is the relaxation time function, X is the relative contribution of sets of protons, T_2 is the water proton relaxation time and i is the number of contributing components. The T_2 signal can be equipped with a mono-exponential model to determine if it comprises one or more components. If more than one component is present, fitting a mono-exponential function will result in an incorrect T_2 value. The number can be calculated by plotting the natural logarithm of the signal $N(t)$ against time, after which it is simple to determine if the function is bi-exponential or tri-exponential, as the resulting plot will have distinct linear regions of a varying slope. This model, however, requires a high signal-to-noise ratio for optimal accuracy. The signal-to-noise ratio is a critical criterion for correct integrations and one of the most effective ways to evaluate an NMR spectrometer's sensitivity. A higher SNR specification indicates that the instrument is more sensitive in general. A line broadening equal to the peak width at half-height is used to achieve an optimal SNR for any sample. The SNR is calculated using equation (6.5).

$$SNR = \frac{2.5L}{N_{pp}}, \tag{6.5}$$

where L is the height of the chosen peak and N_{pp} is the peak-to-peak noise.

When SNR is poor, the fit to the bi-exponential model's four parameters (X_1, X_2, T_{21}, T_{22}) or the tri-exponential model's six parameters ($X_1, X_2, X_3, T_{21}, T_{22}, T_{23}$) become uncertain, lowering accuracy and precision. Furthermore, the relaxation periods should be within the same order of magnitude, and the population fractions should not be less than 15%. It's likely that many more components are present at times. In such instances, using a statistical test to decide whether or not additional works should be done to the fit is the best option.

In determining the cellular water environments in fruits and vegetables using NMR, the relative contribution X_i can be classified according to the pore size, water content, membrane permeability and proton density within the sample to deduce free and bound water environments from T_2^i signal strength [21]. T_2 relaxation is also influenced by water mobility at the molecular level. T_2 is inversely proportional to rotational motion since water mobility is characterised as the ability of water molecules to rotate freely as well as move spatially. Water in a restrictive environment has been shown to have limited relaxation periods, and T_2 is also short when water is actively interacting with macromolecules [22]. The cell wall of plant-based food is mainly made up of solid matter with very little water content. As a result, this structure acts to limit the translational motion and increase the water's correlation time. As a result, the T_2 portion that is short is most likely related to cell wall water or SBW. Furthermore, research shows that the majority of water is found in the intracellular spaces of the food cell, which serve as a water reservoir with a primary water–water interface [23], where the water protons relax slowly, resulting in a longer T_2 relaxation. The longest T_2 component had the highest water fraction of all the T_2 components installed. It can be compared to intracellular water or loosely bound water (LBW) based on the water content of the sample. The last part (medium) is then linked to the free water (FW).

6.5 CURRENT APPLICATION STATUS OF NMR IN BIOLOGICAL TISSUE

NMR spectroscopy is one of the most effective analytical techniques in biology for determining the composition of biological tissues. Water compartmentation in a variety of animal tissues including lung, brain, liver and red blood cells has recently been demonstrated using NMR [24]. T_2 relaxation theories have been extended to the investigation of sugar content in fruit tissue, in determining the quality and the maturity of fruit and vegetables [25]. However, there is a scarcity of literature explaining how to quantify different water environments (e.g. free and bound water) in plant-based foods. Hills and Remigereau [26] looked at T_2 relaxation periods to measure the movement of various types of water in the apple tissue during drying and freezing, but they didn't calculate the types of water in the tissue. Gonzalez et al. [27] investigated the impact of high pressure on the integrity of cell membranes during onion thermal processing. They did not report the data on the proportion of FW, LBW and SBW in the onion tissue when investigating the improvement in T_2 with processing time. T_2 relaxometry has been used in previous NMR studies to research postharvest improvements for quality assurance purposes. Halder et al. [28] examined intracellular water in various plant-based food materials using the BIA process. They have shown the presence of approximately 78–96% water in the intracellular environment, with the amount varying depending on the type of fruit and vegetables. However, they did not focus on the proportion of FW and SBW present in the plant-based food material. Water is found in three major cellular environments in the plant-based food material. Water is distributed differently in different cellular settings.

The structure and properties of food materials have a relationship with the proportion of water in the cells. Different forms of food materials contain different quantities of FW, LBW and SBW due to the structural heterogeneity and diversity of food.

Recently Khan et al. [21] investigated FW, LBW and SBW for various plant-based food materials, as presented in Figures 6.1 and 6.2. According to their findings, different fruit materials contain 80–90% LBW, 2–5% SBW and 10–20% FW, depending on the food structure and physical characteristics. They also found that there is no impact of maturity on cellular water distribution in plant-based food cells. It can be seen that the NMR technology was able to determine the exact proportions of water in different cellular environments.

6.6 APPLICATION OF NMR IN ONLINE MONITORING FOOD QUALITY

Conventional analytical methods are difficult to use for online monitoring in food processing as these are lengthy processes and destructive in nature. Food safety has been assessed using a range of non-invasive detection methods, including light transmission, X-ray transmission and ultrasonic. However, it was difficult to provide precise details about complex food properties using these approaches [16]. NMR has recently been used to accomplish online food quality monitoring. Gudjonsdottir et al. [11], for example, used LF-NMR to investigate online control during shrimp processing by pre-brining with polyphosphates and freezing. They discovered that LF-NMR can detect changes quickly in the physicochemical properties of shrimp such as muscle pH, protein content, phosphate levels, moisture content and water-holding ability. They reported that in using NMR for online control, proper measurement settings, sample replication, analysing surface size and appropriate probe selection need to be optimised. Lv et al. [29] reported that the moisture content of vegetables was measured online during microwave vacuum drying using the LF-NMR process. They discovered that a single mathematical model could not match the relationship between signal amplitude A_{21}, A_{22}, A_{23} and A_2 and moisture content of fresh vegetables (mushroom, carrot, potato, lotus, edamame and corn). They also stated that online measuring performance should be further improved, which would necessitate further research before being used in the industry. As a result, NMR has a lot of potential for online food processing tracking.

6.7 CONCLUSION

Foods can be analysed using NMR and multivariate analysis. The method can be used to examine the water, fat and protein content of various foods. NMR can also detect physical changes in foods, such as colour, texture, water activity and pH. NMR has the ability to provide quantitative data on the chemical

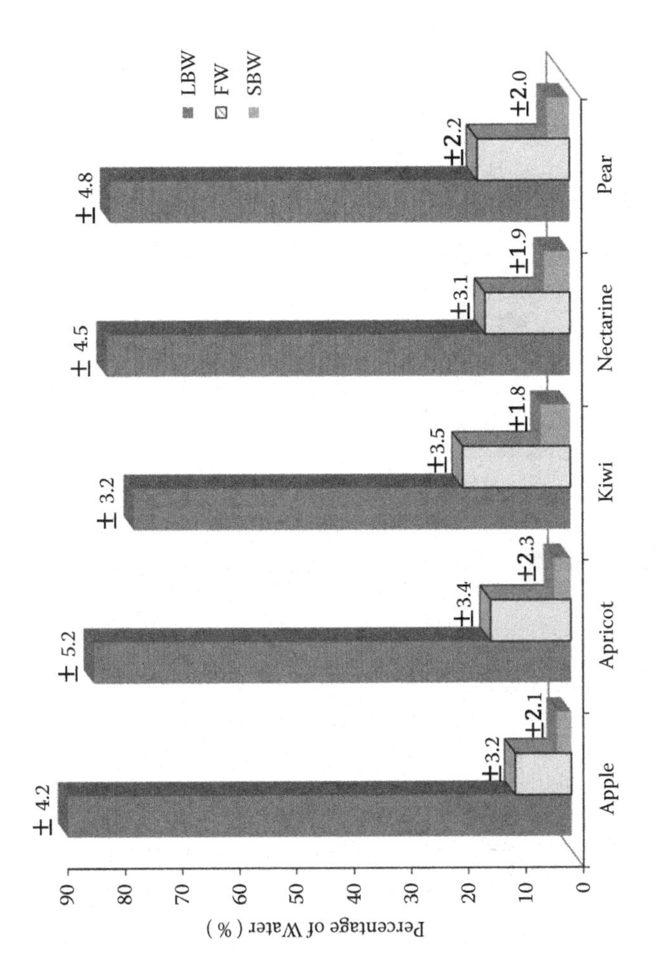

FIGURE 6.1 The percentage of different types of water in various fruits.

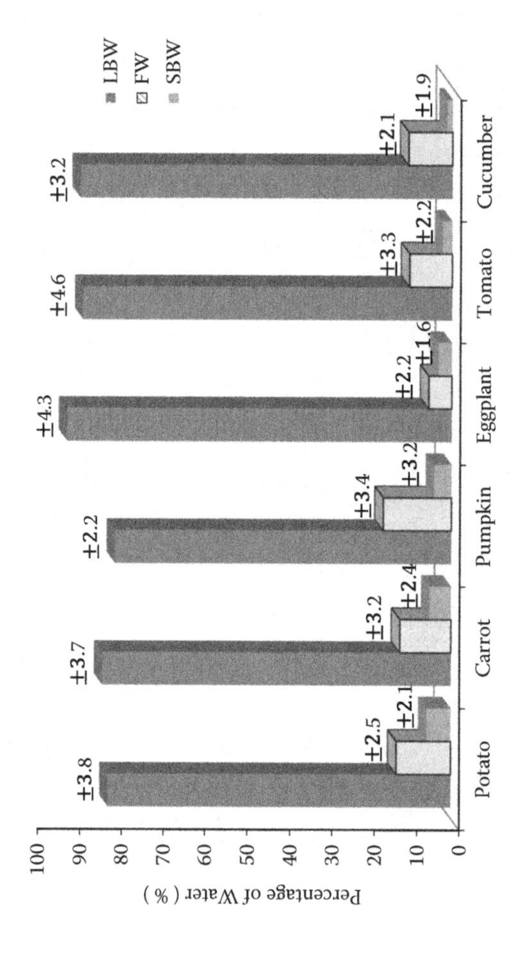

FIGURE 6.2 The proportion of various types of water in various vegetables.

composition of foods. Based on the microbiological quality analysis to assess food quality deterioration, a relationship between NMR parameters and microbiological growth can be established. However, there are still some disadvantages that need more study to effectively apply NMR in the industry.

REFERENCE

1. Kumar, C., et al., *Investigation of intermittent microwave convective drying (IMCD) of food materials by a coupled 3D electromagnetics and multiphase model*. Drying Technology, 2018. **36**(6): pp. 736–750.
2. Duc Pham, N., et al., *Quality of plant-based food materials and its prediction during intermittent drying*. Critical Reviews in Food Science and Nutrition, 2019. **59**(8): pp. 1197–1211.
3. Abesinghe, A., et al., *Effects of ultrasound on the fermentation profile of fermented milk products incorporated with lactic acid bacteria*. International Dairy Journal, 2019. **90**: pp. 1–14.
4. Kumar, C., et al., *A porous media transport model for apple drying*. Biosystems Engineering, 2018. **176**: pp. 12–25.
5. Khan, M.I.H. and M. Karim, *Cellular water distribution, transport, and its investigation methods for plant-based food material*. Food Research International, 2017. **99**: pp. 1–14.
6. Rahman, M.M., et al., *Multi-scale model of food drying: Current status and challenges*. Critical Reviews in Food Science and Nutrition, 2018. **58**(5): pp. 858–876.
7. Rahman, M.M., M.U. Joardder, and A. Karim, *Non-destructive investigation of cellular level moisture distribution and morphological changes during drying of a plant-based food material*. Biosystems Engineering, 2018. **169**: pp. 126–138.
8. Joardder, M.U., et al., *A micro-level investigation of the solid displacement method for porosity determination of dried food*. Journal of Food Engineering, 2015. **166**: pp. 156–164.
9. Mahiuddin, M., et al., *Development of fractional viscoelastic model for characterizing viscoelastic properties of food material during drying*. Food Bioscience, 2018. **23**: pp. 45–53.
10. Rahman, M.M., et al., *A micro-level transport model for plant-based food materials during drying*. Chemical Engineering Science, 2018. **187**: pp. 1–15.
11. Gudjónsdóttir, M., et al., *The effects of pre-salting methods on salt and water distribution of heavily salted cod, as analyzed by 1H and 23Na MRI, 23Na NMR, low-field NMR and physicochemical analysis*. Food Chemistry, 2015. **188**: pp. 664–672.
12. Zhu, F., *NMR spectroscopy of starch systems*. Food Hydrocolloids, 2017. **63**: pp. 611–624.
13. Chen, F.L., Y.M. Wei, and B. Zhang, *Characterization of water state and distribution in textured soybean protein using DSC and NMR*. Journal of Food Engineering, 2010. **100**(3): pp. 522–526.
14. Youssouf, L., et al., *Ultrasound-assisted extraction and structural characterization by NMR of alginates and carrageenans from seaweeds*. Carbohydrate Polymers, 2017. **166**: pp. 55–63.
15. Kirtil, E. and M.H. Oztop, *1H nuclear magnetic resonance relaxometry and magnetic resonance imaging and applications in food science and processing*. Food Engineering Reviews, 2016. **8**(1): pp. 1–22.

16. Marcone, M.F., et al., *Diverse food-based applications of nuclear magnetic resonance (NMR) technology*. Food Research International, 2013. **51**(2): pp. 729–747.
17. Mariette, F., *Investigations of food colloids by NMR and MRI*. Current Opinion in Colloid & Interface Science, 2009. **14**(3): pp. 203–211.
18. Hanson, L.G., *Is quantum mechanics necessary for understanding magnetic resonance?* Concepts in Magnetic Resonance Part A: An Educational Journal, 2008. **32**(5): pp. 329–340.
19. Derome, A.E., *Modern NMR techniques for chemistry research*. 2013, Elsevier.
20. Westbrook, C. and J. Talbot, *MRI in Practice*. 2018, John Wiley & Sons.
21. Khan, M.I.H., et al., *Investigation of bound and free water in plant-based food material using NMR T2 relaxometry*. Innovative Food Science & Emerging Technologies, 2016. **38**: p. 252–261.
22. Rondeau-Mouro, C., et al., *Assessment of TD-NMR and quantitative MRI methods to investigate the apple transformation processes used in the cider-making technology*. In: Francesco, C., Luca, L., Peter, S.B. (eds). Magnetic Resonance in Food Science, 2015. pp. 127–140. https://doi.org/10.1039/9781782622741-00127
23. Joardder, M.U., C. Kumar, and M. Karim, *Food structure: Its formation and relationships with other properties*. Critical Reviews in Food Science and Nutrition, 2017. **57**(6): pp. 1190–1205.
24. Sedin, G., et al., *Lung water and proton magnetic resonance relaxation in preterm and term rabbit pups: their relation to tissue hyaluronan*. Pediatric Research, 2000. **48**(4): pp. 554–559.
25. Delgado-Goñi, T., et al., *Assessment of a 1 H high-resolution magic angle spinning NMR spectroscopy procedure for free sugars quantification in intact plant tissue*. Planta, 2013. **238**(2): pp. 397–413.
26. Hills, B.P. and B. Remigereau, *NMR studies of changes in subcellular water compartmentation in parenchyma apple tissue during drying and freezing*. International Journal of Food Science & Technology, 1997. **32**(1): pp. 51–61.
27. Gonzalez, M.E., et al., *1H-NMR study of the impact of high pressure and thermal processing on cell membrane integrity of onions*. Journal of Food Science, 2010. **75**(7): pp. E417–E425.
28. Halder, A., A.K. Datta, and R.M. Spanswick, *Water transport in cellular tissues during thermal processing*. AIChE Journal, 2011. **57**(9): pp. 2574–2588.
29. Lv, W., et al., *Smart NMR method of measurement of moisture content of vegetables during microwave vacuum drying*. Food and Bioprocess Technology, 2017. **10**(12): pp. 2251–2260.

7 Near-infrared (NIR) Spectroscopy for Food Processing Applications

7.1 INTRODUCTION

The processing of agricultural products demands special attention, as these are considered important sources of bioactive compounds and minerals essential for humankind [1,2]. Post-harvest losses of fruits and vegetables, mostly due to the lack of proper processing, are estimated to be about a third of the total global production [3,4]. Moreover, there has been a growing demand for healthy processed foods [5]. Food protection and human health have become increasingly important in recent years, and therefore, high-quality dried foods are gaining commercial importance. Industries and researchers are increasingly looking at the source of quality deterioration rather than finding the poor quality foods and separating them from good foods. Micro-level features of the food items dominate the macro-level changes [6,7], and therefore it is critical to detect the quality problems when it starts at a cell level [8]. Cellular structural and water distribution information is critical for the early detection of a quality problem [9,10].

Traditional quality detection methods are lengthy and destructive. In the recent times, rapid and non-destructive techniques for evaluating food hazards, food authentication and traceability have made significant progress. NIR spectroscopy is a non-destructive analytical technique that offers chemical and physical knowledge about any complex structure in a short amount of time. The widespread adoption of NIR spectroscopy and imaging has been aided by recent advancements in instrumentation. Using multivariate regression to perform the quantitative and qualitative analysis is time consuming. As a result, NIR spectroscopy has gained widespread acceptance in a wide variety of applications, including the food and pharmaceutical industries, petroleum, pharmaceutics and agriculture.

NIR reflectance spectroscopy has developed itself as an effective analytical technique in the fields of food and agriculture. In this method, before any measurements could be made, the samples had to be ground into powder form. It is, however, faster and simpler nowadays because it does not require any pre-treatments or special sample preparations. Because of advances in instrumentation technology, the availability of open-source software and improvements in photonics materials, NIR technology would become a viable alternative to traditional chemical test methods. The ability of

DOI: 10.1201/9781003047018-7

spectrophotometers to immediately record spectra makes this device suitable for online research.

The use of NIR spectroscopy techniques in food quality analysis is becoming more popular. There is a lot of research going on right now to see whether NIR spectroscopy can be used to detect harmful pathogens in milk, other dairy products and animal products (meat and eggs). The methodology of NIR spectroscopy, as well as its various applications in food processing, is discussed in this chapter.

7.2 HISTORY OF NIR

Friedrich Wilhelm Herschel discovered near-infrared radiation in 1800, which had a wavelength spectrum of 780–2500 nm by definition. The incident radiation may be mirrored, absorbed or transmitted when it reaches a sample. The relative value of each phenomenon is determined by the sample's chemical composition and physical parameters [11].

While Herschel discovered light in the near-infrared region in 1800, spectroscopy was overlooked until the first half of the twentieth century because it was considered to lack analytical interest. Although the first application of NIR was published in the 1950s, a group of researchers applied it to the study of agricultural food samples in the 1970s. The expansion of this technique in various fields was helped by the production of equipment with improved electronic and optical components, as well as the implementation of computers capable of effectively processing the information found in NIR spectra.

The advantages of NIR spectroscopy over other instrumental techniques are what piqued people's interest. It can record spectra for solids and liquids without any pre-treatment, apply continuous methodologies, provide spectra immediately and predict physical and chemical parameters using a single spectrum. As a result of these characteristics, it is capable of easy and fast characterisation of samples, which is appealing.

7.3 NIR METHODS

Frequency, wavelength and energy are the characteristics of electromagnetic energy, which spans a wide spectrum. The visible light spectrum has wavelengths of 400–700 nm. The electromagnetic spectrum's next range is infrared radiation. NIR and far infrared (FIR) wavelengths have been separated. Near-infrared radiation matches the visible red spectrum and has a wavelength range of 750–3000 nm. Infrared radiation is generated by molecular motions, transformations or rotation of solids, gases and liquids in theory. As molecules absorb radiation energy, they are excited, causing them to vibrate and produce an infrared absorption spectrum. Dependent on the molecular structure, the frequency or wavelength of the molecular vibration has a primary peak and a sequence of harmonics or overtones. The shape of any material's spectrum is determined by these vibrational characteristics. The fundamental bands' positions in the IR region are well established. By dividing the wavelength of the fundamental vibration by 2, 3 and 4 for the first, second and third overtones, respectively, the approximate location of the overtones can be determined. In the NIR region, overtones and combination bands of fundamental bands from the IR region can be found. These overtones and hybrid bands are usually weaker than the fundamental band. Organic materials' spectral responses are also more pronounced in the mid-IR field. The positions of the functional classes, on the other hand, are well differentiated in the NIR field. As a result, NIR spectra are better suited to quantitative rather than qualitative material composition analysis. NIR spectroscopy is based on the interaction of electromagnetic radiation with matter and the energy transfer that results.

The vibrational response of O–H, C–H, C–O and N–H molecular bonds is the primary source of electromagnetic response in the NIR field. In particular, hydrogen bonds have higher spectral stability in the near-infrared region, which aids in food quality and composition analysis [12].

The energy absorption bands in the NIR region are well aligned with functional group bonds. Depending on the overtone of the absorption bands in the NIR zone, the strength of this absorption energy may increase or decrease. NIR bands, on the other hand, have a much lower absorptivity than mid-infrared bands, allowing NIR radiation to penetrate deeper into a sample and provide a better composition analysis. Various chemical bonds are contained in water, ethanol, sugars (fructose and glucose), organic acids, phenolic compounds and food oxidative products found in food. The ability to absorb energy in the NIR spectrum has increased the likelihood of using NIR spectroscopy to measure food quality.

7.4 WORKING PRINCIPLE OF NIR SPECTROSCOPY

The NIR region of the electromagnetic spectrum is described by the American Society of Testing and Materials (ASTM) as the wavelength range of 780–2500 nm, which corresponds to the wave number range of 12,820–4,000 cm^{-1}. Overtones and combinations of fundamental vibrations of –CH, –NH, –OH and –SH functional groups are the most prominent absorption bands in the NIR field. The NIR spectrum and chemical composition response at effective wavelengths are shown in Figure 7.1.

TRANSMITTANCE

As shown in Figure 7.2(a), light in the NIR range falls from the source onto the sample, and as it passes through the sample, the sample absorbs wavelength-specific light, which can then be measured from the other side of the source to provide a result analysis. It is used for high water content samples and is connected to 1,100–1,800 nm wavelengths.

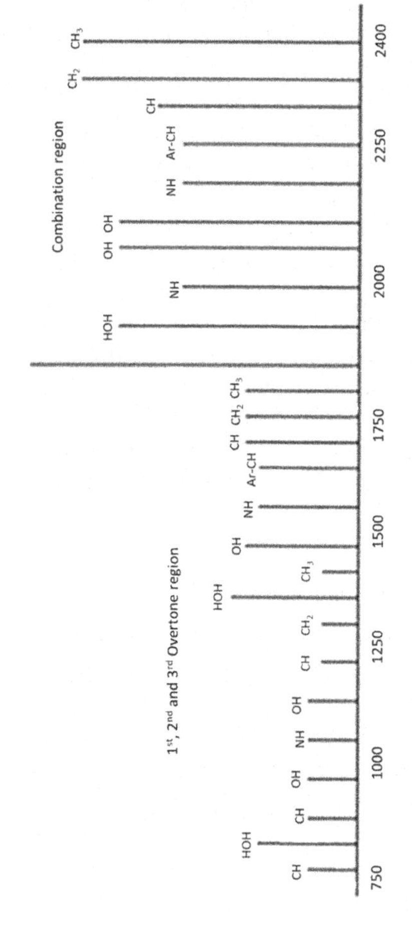

FIGURE 7.1 Constituent response at different wavelengths.

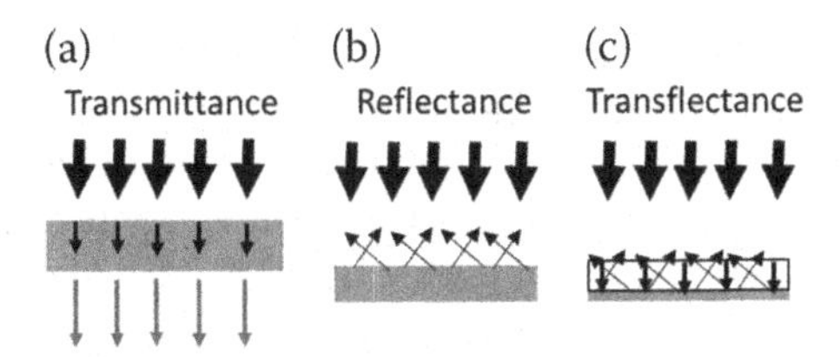

FIGURE 7.2 Transmittance, reflectance and transflectance.

Reflectance

In the case of these occurrences, the light emitted by the sample's reflective properties will be used to predict the sample analysis, shown in Figure 7.2(b). This phenomenon occurs between 1,800 and 2,500 nm in wavelength and is used to study earth samples and solids.

Transflectance

Figure 7.2(c) depicts a transflectance phenomenon, which is the combination of transmittance and reflectance. As a result, sample interpretation predictions would be based on both reflected and transmitted light from the sample. Transflectance occurs at wavelengths of 800–1,100 nm and is used to analyse slurries, grains and dense samples.

7.5 DIVERSE APPLICATIONS OF NIR SPECTROSCOPY IN FOOD SCIENCE

To clarify the research patterns in the food industry, a detailed literature survey has been presented here. Elke Anklam et al.'s [13] research article "Analytical Methods for Identification and Determination of Genetically Modified Organisms in Agriculture Crops and Plant-Derived Food Products" highlights the value of NIR spectroscopy for detecting genetically modified organisms (GMOs). The possible economic effect of GMO commingling in the supply and marketing chain has also been studied, establishing the utility of NIR reflectance spectroscopic analysis in GMO detection.

Haiqing Yang's research article published in 2011 "Remote Sensing Technique for Predicting Tomato Harvest Time" is based on various growth stages and harvest times of fruits and vegetables for horticulture automation [14]. The study has proposed a growing stage (GS) index for predicting the harvest time of three different tomato cultivars using NIR spectral response. The GS index was determined as a result of measuring the NIR spectral response. It was found that visible–near-infrared (VIS–NIR) spectroscopy, combined with optimised PLSR models for GS prediction, could be successfully used as a remote sensing technique for predicting tomato harvest time, enabling autonomous fruit picking robots to be implemented.

Another research conducted by Hao Lin in 2011 [15] at the School of Food and Biological Engineering, Jiangsu University, China, on "Freshness Calculation of Eggs Using NIR Spectroscopy and Multivariate Data Analysis" revealed the importance of NIR techniques for food analysis at various levels. Furthermore, Linda M. Reid, Colm P.O'Donnell and others [16] have concentrated on "Recent Technical Developments for the Determination of Food Authenticity," with particular attention to the European system for producing and preserving food in 2006. The relative capacity of varying technologies for verifying food authenticity and consistency has been briefly explored. The authors have also looked into the use of NIR spectroscopy to assess the authenticity of food. The importance of continuing to develop applications for more proven techniques based on NIR techniques with chemometrics analysis was determined.

In an article titled "Non-Destructive Measurement of Fruit and Vegetable Quality by Means of NIR Spectroscopy," Bart M. Nicolai et al. [17] from the Flanders Centre of Postharvest Technology in Belgium demonstrated the efficacy of the NIR technique in 2007. The study's aim was to determine the efficacy of online systems for grading fruits and vegetables, which would then add market value to the items. Premium buyers who are willing to pay premium rates for extra sweet fruit, for example, would assist them in selecting the best commodity, allowing the auction house to provide grading lines for their produce.

Another research article was published by C. Nick Pace et al. [18] in 1995 on "How to Calculate and Predict the Molar Absorption Coefficient of Protein." Texas A&M University's Department of Medical Biochemistry and Genetics has pioneered the content determination for multivariate content analysis in food engineering. A summary of the latest food industry research pattern using NIR reflectance spectroscopy represents different aspects of fruits, vegetables and crops. With advances in NIR reflectance spectrometry, NIR will be a perfect solution for the quality-based ranking of both fruits and vegetables. Major research in agriculture science has begun and is picking up pace in terms of food safety. In the coming years, NIR spectroscopy will be the most widely accepted and favoured alternative to traditional chemical methods. NIR spectroscopy is more widely accepted than traditional methods because it is a non-invasive approach that allows for fast study.

7.5.1 Determination of Moisture Content (MC) by NIR Spectroscopy

Moisture content is one of the most significant consistency parameters for dried food commodities. To prevent the growth of microorganisms and moulds, fruits are usually dried to a moisture content of 16–25%. Traditional methods for measuring moisture content include oven drying, Karl-Fischer titration and distillation, all of which are sample-invasive and time consuming. Since water's O–H functional group absorbs energy at particular wavebands during electromagnetic radiation, spectroscopic methods have been used to determine the moisture content [19]. Due to their spot-detecting nature, current spectrometers are much more useful in providing average information for the entire sample and thus are not appropriate for measuring the irregular moisture distribution of the sample.

NIR was first used 33 years ago as part of an effort to create a technique for rapid moisture analysis of grains [20]. Since then, NIR has been used to examine a number of food products, including protein and moisture in wheat, various constituents in milk and milk products, and fat, protein, and moisture in fresh meat [21]. NIR has also been used to determine the moisture content of a variety of fruits and vegetables.

Hyperspectral imaging (HSI) is now widely used as a quick, non-invasive and environmental-friendly detecting method for food safety and quality assessment, especially for meat products [22]. Since the HSI method combines computer vision and spectroscopic techniques into one system, it shows target details of the sample in one image using a built model. HSI has been effective in identifying the surface or internal defects of fruits. Meanwhile, HSI has proven to be effective in detecting fruit fly/insect infestations or faecal contamination. Chemical components such as water content, soluble solid content, acidity and phenolics can also be expected. The use of HSI to assess the changes in MC in strawberries and bananas during the ripening process was investigated [23]. Over the entire spectral range of 400–1,000 nm, it was discovered that the mean relative reflectance of overripe strawberries (with a higher moisture content) was lower than that of unripe strawberries. Similarly, the moisture content of the banana pulp increased with maturity, and using selected wavebands in a multiple linear regression (MLR) model, the MC prediction accuracy reached R^2 of 0.87. Using Vis–NIR hyperspectral imaging, the moisture content of sliced mushrooms during storage and whole mushrooms dried by a convective air dryer [24] was successfully predicted. Soybean moistures at various drying times (from 10 to 80 minutes in a 10-minute interval) have also been investigated using microwave-assisted pulse-spouted bed vacuum-drying.

After drying, the dried samples were removed from the desiccator and scanned with hyperspectral images. The research used two line-scan reflectance hyperspectral imaging systems. The first was carried out in the Vis–NIR spectral range (400–1,000 nm with a spectral resolution of 5 nm), while the second was carried out in the NIR spectral range (880–1,720 nm with a spectral resolution of 7 nm). The device automatically calibrated the obtained raw images using a white and a dark reference image.

The white reference image was created by scanning a uniform white ceramic tile, and the dark reference was created by covering the lens with the light turned off. The reference moisture contents of the samples at different drying stages were calculated using a thermo-gravimetric method in a convective hot-air oven at 105°C for 24 hours after image acquisition. Weighing samples before and after oven drying was done separately. The moisture content was calculated on a wet basis (w.b.) and expressed as follows:

$$\text{Moisture content (\% w.b.)} = \frac{W_1 - W_2}{W_1} \times 100 \quad (7.1)$$

where w_1 and w_2 were sample weights before and after oven drying.

7.5.2 On-line Properties Measurement by NIR Spectroscopy

The growing demand for food quality assurance necessitates the development of sophisticated analytical methods for objective quality control. Traditional analytical approaches, on the other hand, are time consuming, labour-intensive and costly. Vibrational techniques, such as NIR spectroscopy, provide a clear, fast and cost-effective solution. In- and on-line applications are possible with NIR spectroscopy since it makes measurements without prior sample preparation. As a result, this technique satisfies the criteria for continuous quality control and process monitoring in industrial applications. However, since the method is based on indirect measurements, the resulting spectra are highly convoluted and broad, making them nearly impossible to interpret with the naked eye. As a result, in order to obtain analytical information from the corresponding spectra, NIR spectroscopy involves calibration using mathematical and statistical methods (chemometrics).

7.5.3 Screening Out Adulterated Samples from the Food Supply Chain (On-line Quality Check)

Spices are used to improve the organoleptic qualities of food and culinary dishes, making them more appealing to consumers. The use of illegal, low-cost colorants may be lucrative in the food supply chain, but it puts human health at risk. NIR spectroscopy, combined with chemometrics, can be used as a fast, simple, non-destructive and low-cost screening method for detecting adulterants in food samples. NIR spectroscopy has also been used to screen out the adulterated samples within the food supply chain. Thereby, this technique has been beneficial in the on-line quality checking of the food products.

Several analytical methodologies for evaluating Sudan and Para-Red dyes in food products have been established, with chromatographic techniques being the most widely used with various detection methods [25]. Alternatively, spectrometric approaches such as UV–visible and fluorescence can be created to achieve these goals. Emerging vibrational spectroscopy techniques, in particular, are molecular fingerprinting methods that are becoming increasingly common in food screening analysis [26] due to their speed, ease of use in industrial applications and ease of use in routine work. Raman, Fourier transform infrared (FT-IR) and near-infrared (NIR) spectroscopies have been proposed to evaluate Sudan dyes in food in this context. There are only a few Surface Enhanced Raman Spectroscopy (SERS) methods available in Raman. Haughey et al. [27] have proposed using traditional Raman and NIR spectroscopy to detect Sudan I dye in chilli powder. FT-IR was also used to classify Sudan I dye in paprika powder [28]. These techniques have the potential to be portable for in-situ analysis, but NIR has a few advantages over IR: it penetrates far deeper into an intact food sample, allowing for the analysis of food products through plastic or glass packaging materials; water does not cause interference; and no additional accessories like attenuated total reflection (ATR) are needed. All of these advantages make NIR an excellent method for regular and repeated in-situ analyses.

NIR with multivariate analysis has been used to detect adulterations in various food and agricultural products for a rapid, bulk and high-throughput food screening analysis, in addition to the aforementioned studies.

7.5.4 Non-destructive Estimation of Maturity of Fruits

Apart from the appearance provided to the fruit by scale, colour, form and surface defects, total soluble solids (TSS), which is directly related to sugar content and titratable acidity (TA), are important indicators of internal fruit quality. Non-destructive optical methods focused on visible/near-infrared spectroscopy (VIS/NIRS) have been tested for the non-destructive estimation of internal starch, soluble solids content, oil content, water content, dry-matter content, acidity, firmness, stiffness factor and other physiological properties of a variety of fruit and vegetable products, including citrus [29], mandarin [30], tomato [31], mango [32] and kiwifruit[33]. TSS and TA are the key internal consistency parameters used all over the world in citrus fruit juice content. Managers of packinghouses take representative samples of fruit to measure internal consistency before shipment, but where the acceptable minimum level of TSS is not met or TA reaches the highest tolerated level, there is controversy. Even within the same cultivar, the TSS and TA content can differ greatly depending on a variety of factors: the duration of the blooming season, which results in the setting of fruits of varying ages in the same tree that are difficult to distinguish at the harvest time, the location of the fruits in the canopy and the type of inflorescences that carry the fruit. As a result, a reliable non-destructive analysis is needed to select fruits based on their true and individual quality characteristics, as well as to ensure that they meet the requisite internal quality requirements.

The majority of instrumental techniques for calculating such properties, on the other hand, are destructive and time consuming. As a result, a non-destructive method based on a large number of VIS/NIRS may estimate both soluble solid content and acidity, which are significant internal quality features. The advancement of non-destructive analysis for TSS and TA determination represents a significant advance in the citrus fruit industry, enabling fruit quality classification not only on the basis of visual appearance but also in relation to gustative characteristics. This could lead to improvements in farming methods and techniques, as farmers become aware of the prospect of receiving a higher price for the fruit's internal quality rather than just its aesthetic qualities.

7.6 ADVANTAGES AND LIMITATIONS OF NIR SPECTROSCOPY

The benefits and drawbacks of NIR spectroscopy are summarised below.

7.6.1 Advantages of NIR Spectroscopy

- It is a non-destructive, non-invasive procedure.
- It necessitates little to no sample planning. If a suitable instrument is used, solid samples may be directly tested with minimal to no pre-treatment.
- Measurement and distribution of results are almost instantaneous. The real-time extraction of analytical information from samples has been made, possible thanks to significant advancements in NIR equipment and chemometrics used in conjunction with computers.
- No reagents or materials are needed to prepare samples, and the automation of the procedure contributes to improved throughput, which decreases analytical costs and amortisation time.
- Multiple analytes can be calculated at the same time using a single spectrum. Non-chemical (physical) parameters can be determined using this technique. Indeed, the effect of certain parameters on the NIR spectrum allows for the fast determination of properties including density, viscosity and particle size.
- NIR instrumentation is best suited for process control at the production level due to the high strength of optical materials and the robustness of NIR equipment, which in most cases has no moving parts.
- Fibre-optic offers stable, powerful sensors for process control at-line, on-line and in-line research.
- The precision of NIR spectroscopic results is comparable to that of other analytical techniques, and their precision is typically higher because no sample treatment is needed.

7.6.2 Disadvantages of NIR Spectroscopy

- Since NIR measurements are not very selective, chemometrics techniques must be used to model data from which relevant information can be extracted.
- There are no reliable models that account for NIR light's interaction with matter. As a consequence, in many instances, calibration is strictly empirical.
- Accurate and reliable calibration models are difficult to come by because they require a large number of samples to account for all differences in physical and/or chemical properties.
- The need to account for sample physical and chemical variability in calibration necessitates the use of as many different calibration models as there are sample forms, resulting in multiple models per analyte.

- Since the methodology isn't really sensitive, it's generally limited to major components.
- Since NIR spectroscopy is a relative technique, building models with it necessitates a prior knowledge of the target parameter's value, which must be calculated using a reference component.
- Constructing NIR models necessitates a large expenditure, which can be amortised by moving calibrations from the master equipment to a variety of slaves.

7.8 CONCLUSION

In order to explore the enormous potential of NIR chemometrics with special regard to the food and agriculture industries, a review of the possible uses of NIR in food science has been presented and trending research in the field of near-infrared spectroscopy has been identified. NIR chemometrics may be preferable to traditional chemical methods for the study of multivariate material samples because it is a non-invasive and non-destructive technique. In the past few decades, tremendous technical advances have occurred, resulting in NIR spectroscopy and imaging being more commonly used in the food industry to satisfy the urgent need for rapid and non-destructive food testing. On-line/real-time applications are projected to be a major growth area in the foreseeable future, and this technology will certainly play an indispensable role in food safety assessment.

REFERENCES

1. Kumar, C., et al., *A porous media transport model for apple drying*. Biosystems Engineering, 2018. **176**: pp. 12–25.
2. Abesinghe, A., et al., *Effects of ultrasound on the fermentation profile of fermented milk products incorporated with lactic acid bacteria*. International Dairy Journal, 2019. **90**: pp. 1–14.
3. Joardder, M.U., R.J. Brown, and A. Karim, *Prediction of shrinkage and porosity during drying: Considering both material properties and process conditions* In International Congress on Engineering and Food, 2015: pp. 1–1.
4. Khan, M.I.H., et al., *Fundamental understanding of cellular water transport process in bio-food material during drying*. Scientific Reports, 2018. **8**(1): pp. 15191.
5. Duc Pham, N., et al., *Quality of plant-based food materials and its prediction during intermittent drying*. Critical Reviews in Food Science and Nutrition, 2019. **59**(8): pp. 1197–1211.
6. Rahman, M.M., M.U. Joardder, and A. Karim, *Non-destructive investigation of cellular level moisture distribution and morphological changes during drying of a plant-based food material*. Biosystems Engineering, 2018. **169**: pp. 126–138.
7. Rahman, M.M., et al., *Multi-scale model of food drying: Current status and challenges*. Critical Reviews in Food Science and Nutrition, 2018. **58**(5): pp. 858–876.

8. Mahiuddin, M., et al., *Development of fractional viscoelastic model for characterizing viscoelapstic properties of food material during drying*. Food Bioscience, 2018. **23**: pp. 45–53.
9. Khan, M.I.H. and M. Karim, *Cellular water distribution, transport, and its investigation methods for plant-based food material*. Food Research International, 2017. **99**: pp. 1–14.
10. Rahman, M.M., et al., *A micro-level transport model for plant-based food materials during drying*. Chemical Engineering Science, 2018. **187**: pp. 1–15.
11. Blanco, M. and I. Villarroya, *NIR spectroscopy: a rapid-response analytical tool*. TrAC Trends in Analytical Chemistry, 2002. **21**(4): pp. 240–250.
12. Shah, M.K., C.B. Bhatt, and J.B. Dave, *NIR spectroscopy: Technology ready for food industries applications*. International Journal of Applied and Natural Sciences, 2016. **5**(1): pp. 129–138.
13. Anklam, E., et al., *Analytical methods for detection and determination of genetically modified organisms in agricultural crops and plant-derived food products*. European Food Research and Technology, 2002. **214**(1): pp. 3–26.
14. Yang, H., *Remote sensing technique for predicting harvest time of tomatoes*. Procedia Environmental Sciences, 2011. **10**: pp. 666–671.
15. Lin, H., et al., *Freshness measurement of eggs using near infrared (NIR) spectroscopy and multivariate data analysis*. Innovative Food Science & Emerging Technologies, 2011. **12**(2): pp. 182–186.
16. Reid, L.M., C.P. O'donnell, and G. Downey, *Recent technological advances for the determination of food authenticity*. Trends in Food Science & Technology, 2006. **17**(7): pp. 344–353.
17. Nicolai, B.M., et al., *Nondestructive measurement of fruit and vegetable quality by means of NIR spectroscopy: A review*. Postharvest Biology and Technology, 2007. **46**(2): pp. 99–118.
18. Pace, C.N., et al., *How to measure and predict the molar absorption coefficient of a protein*. Protein Science, 1995. **4**(11): pp. 2411–2423.
19. Magwaza, L.S., et al., *NIR spectroscopy applications for internal and external quality analysis of citrus fruit—a review*. Food and Bioprocess Technology, 2012. **5**(2): pp. 425–444.
20. Norris, K.H., *History, present status, and future prospects for NIRS*. In: Creaser, C.S., Davies, A.M.C., editors. Analytical applications of spectroscopy, 1988. pp. 1–7.
21. Kruggel, W.G., et al., *Near-infrared reflectance determination of fat, protein, and moisture in fresh meat*. Journal of the Association of Official Analytical Chemists, 1981. **64**(3): pp. 692–696.
22. Wu, D. and D.-W. Sun, *Potential of time series-hyperspectral imaging (TS-HSI) for non-invasive determination of microbial spoilage of salmon flesh*. Talanta, 2013. **111**: pp. 39–46.
23. Rajkumar, P., et al., *Studies on banana fruit quality and maturity stages using hyperspectral imaging*. Journal of Food Engineering, 2012. **108**(1): pp. 194–200.
24. Taghizadeh, M., A. Gowen, and C.P. O'Donnell, *Prediction of white button mushroom (Agaricus bisporus) moisture content using hyperspectral imaging*. Sensing and Instrumentation for Food Quality and Safety, 2009. **3**(4): pp. 219–226.
25. Reinholds, I., et al., *Analytical techniques combined with chemometrics for authentication and determination of contaminants in condiments: A review*. Journal of Food Composition and Analysis, 2015. **44**: pp. 56–72.

26. Ellis, D.I., et al., *Point-and-shoot: rapid quantitative detection methods for on-site food fraud analysis–moving out of the laboratory and into the food supply chain.* Analytical Methods, 2015. **7**(22): pp. 9401–9414.
27. Haughey, S.A., et al., *The feasibility of using near infrared and Raman spectroscopic techniques to detect fraudulent adulteration of chili powders with Sudan dye.* Food Control, 2015. **48**: pp. 75–83.
28. Lohumi, S., et al., *Quantitative analysis of Sudan dye adulteration in paprika powder using FTIR spectroscopy.* Food Additives & Contaminants: Part A, 2017. **34**(5): pp. 678–686.
29. Zude, M., et al., *NIRS as a tool for precision horticulture in the citrus industry.* Biosystems Engineering, 2008. **99**(3): pp. 455–459.
30. McGlone, V.A., et al., *Internal quality assessment of mandarin fruit by vis/NIR spectroscopy.* Journal of Near Infrared Spectroscopy, 2003. **11**(5): pp. 323–332.
31. Slaughter, D., D. Barrett, and M. Boersig, *Nondestructive determination of soluble solids in tomatoes using near infrared spectroscopy.* Journal of Food Science, 1996. **61**(4): pp. 695–697.
32. Saranwong, S., J. Sornsrivichai, and S. Kawano, *Prediction of ripe-stage eating quality of mango fruit from its harvest quality measured nondestructively by near infrared spectroscopy.* Postharvest Biology and Technology, 2004. **31**(2): pp. 137–145.
33. Osborne, S.D., R. Künnemeyer, and R.B. Jordan, *A low-cost system for the grading of kiwifruit.* Journal of Near Infrared Spectroscopy, 1999. **7**(1): pp. 9–15.

8 Light Microscopy (LM) in Food Processing

8.1 INTRODUCTION

Advances in imaging techniques allow industries and researchers to move from traditional bulk-level inspection to micro-level investigations using images of the food structure [1,2]. Plant-based food materials are considered to have a multi-level architecture, and therefore cellular-level investigation is important [3,4]. Literature suggested that there is a link between the microstructural changes and food quality [5,6]. Imaging techniques are considered to be the most suitable techniques for analysing the food microstructure [7,8]. Until now, the majority of advances in microscopy and imaging techniques are made outside the area of the food science, in areas such as materials science and medicine. In most cases, such methods cannot be directly applicable to the analysis of the food structure [9,10]. They must be adapted because the processing conditions that convert plan-based food materials into processed food result in structural and textural changes, which alter the foods' inherent properties and behaviour. This necessitates the development of suitable methods as well as specialised knowledge. Future advances in this field can be divided into two categories: the use of new equipment designed for other fields and the use of techniques adapted to solve specific food science problems, such as the production of new foods with specific properties and textures or the identification of food defects.

Image data is produced by microscopy (e.g. optical or light microscopy), which are an advancement from the traditional visual inspection of foods that both customers and food producers have done in the past. Microscopy techniques differ in image production process, resolution and the type of signal detected, and each technique provides a specific type of structural details [9]. Foods with similar structures may be loosely grouped as foods with similar textures. Most foods are biologically produced, but they are processed to varying degrees, often to the point that their biological origin is obscured, as in grain versus wheat, meat versus salami or milk versus cheese. Food processing results in changes at the microscopic and molecular stages, such as grain milling, gelatinisation of starch, comminution of meat, heat denaturation of proteins, gelation of milk and proteolysis of proteins. Imaging methods may be used to assess certain morphological and compositional changes [11].

Optical microscopy was first used in food science to detect accidental or intentional food contamination or adulteration. This was accompanied by a fascination with the food microstructure and how it applied to other aspects of a specific food. With the popularisation of food production, it is now important to

DOI: 10.1201/9781003047018-8

understand the processes that lead to the formation of various structures (such as foams, emulsions, dispersions, extrudes and fibres) to produce such structures in newly designed foods and to prevent defects in the foods being produced.

Food structural analysis has progressed exponentially in the 25 years [9]. This progress has been examined by Aguilera and Stanley [12], who have discussed the methods, observations and interpretations. True food structure visualisation is extremely difficult as each stage of preparing a specimen for microscopy changes the food sample in some way [9]. During preparation, inadvertent or intentional removal of water, lipid or other substances affects the sample and the relationships between its components. As a consequence, when drawing conclusions and generalisations from the empirical findings, certain improvements must be taken into account. The best approach is to compare and validate findings by subjecting each food sample to multiple imaging techniques. The strength of imaging techniques lies in their application as part of a larger scheme, in which changes detected at different resolution levels are routinely evaluated.

8.2 METHODS AND MECHANISMS OF LIGHT MICROSCOPY

The most common techniques used are bright-field, polarising and fluorescence microscopy. While manufacturers produce "new and improved" microscopes and other peripheral equipment on a regular basis, the fundamentals of the image forming in these techniques remain the same.

8.2.1 Bright-field Microscopy

Lighting is transmitted sequentially through a condenser, the specimen and the target in traditional bright-field microscopy, resulting in a real image that is inverted and magnified inside the microscope tube. The actual image is then magnified once more by the ocular lens, resulting in either a simulated image that appears to be 25 cm away from the eye or a real image on a photographic film (or video) mounted above the microscope tube. If the specimen is not brightly coloured, contrast must be added to help it stand out. The use of dyes or stains with proven specificity for various components of the specimen is a common way to accomplish this. For the food microscopist, the blue staining of starch with iodine is a long-established method but is still a very useful reaction. Many stains are useful for protein localisation, like Fast Green and Acid Fuchsin. Toluidine Blue O (TBO) is a metachromatic dye that is particularly useful in the analysis of foods: TBO stains pectin-containing plant cell walls (e.g. in fruit and vegetable tissues) pink to purple, while TBO stains lignin-containing plant cell walls (e.g. in vascular tissues) to dark blue. Muscle tissue is light pink, fibroblasts are bluish and elastin fibres are turquoise in meat products stained with TBO. Fats are stained with lipid-soluble dyes like Oil Red O. Altering the direction of light entering the specimen, such as in polarising microscopy, is another way to introduce contrast.

8.2.2 Polarising Microscopy

By inserting two polarisers in the light direction, one between the light source and the specimen and the other between the sample and the observer, a typical bright-field microscope can easily be converted to polarising microscopy. The first polariser produces plane-polarised light, which has rays that vibrate only in one plane, perpendicular to the direction of travel. Polaroid film is a popular and inexpensive way to achieve this effect. Amorphous regions inside the specimen will appear dark if the second polariser, the analyser, is rotated such that the emitted vibration is at right angles to the vibration of the incident light (crossed polar), while crystalline or ordered regions will appear very bright against a dark backdrop. The bright areas are caused by the fact that their constituents have two primary refractive indices and these substances are known as birefringent. Many food ingredients, including starch, fats, plant cell walls, muscle fibres and many flavour and seasoning ingredients, are birefringent.

8.2.3 Fluorescence Microscopy

Light of a short wavelength is absorbed by specific molecules present in the specimen in fluorescence microscopy, and the energy is re-emitted as the light of a longer wavelength and lower intensity (fluorescence) [13]. Instead of transmitted light, the most versatile fluorescence microscopes use incident light or epi-illumination. The objective lens serves as both the lens and the objective lens simultaneously in epi-illumination, removing the need to coordinate the two lenses. A chromatic beam splitter, also known as a dichroic mirror, is a device that is placed between the battle source and the target and reflects the light of shorter wavelengths while transmitting the light of longer wavelengths. As a result, the shorter-wavelength thrilling light is reflected down onto the specimen, while the specimen's longer-wavelength fluorescence is transmitted to the eyepiece. Furthermore, extraneous shorter-wavelength light transmitted from the specimen or the optics is reflected by the dichroic mirror, preventing it from reaching the eyepiece, leaving only the absorbed fluorescence to form the image seen.

Many food products, both plant and animal-based, have a natural fluorescence (auto-fluorescence). Pigments (such as chlorophyll and carotenoids) and both high- and low-molecular mass phenolic compounds are examples of these in plants (e.g. lignin and ferulic acid). Bone and cartilage, collagen, elastin and certain fats are the primary causes of auto-fluorescence in animal tissues. Vitamins, flavourings and seasonings are among the fluorescent ingredients. In addition to auto-fluorescence, a growing number of fluorescent probes designed to impart fluorescence to the component of interest are available. Calcofluor, a fluorescent brightener, is used to localise $\beta(1 \rightarrow 3), (1 \rightarrow 4)$-D-glucan in cereal grains and fluoresces in solution. Others, such as Nile blue, which has a portion that is lipid soluble, need a particular environment, such as the non-polar milieu within a fat droplet, to be fluorescent. Antibodies and lectins labelled with

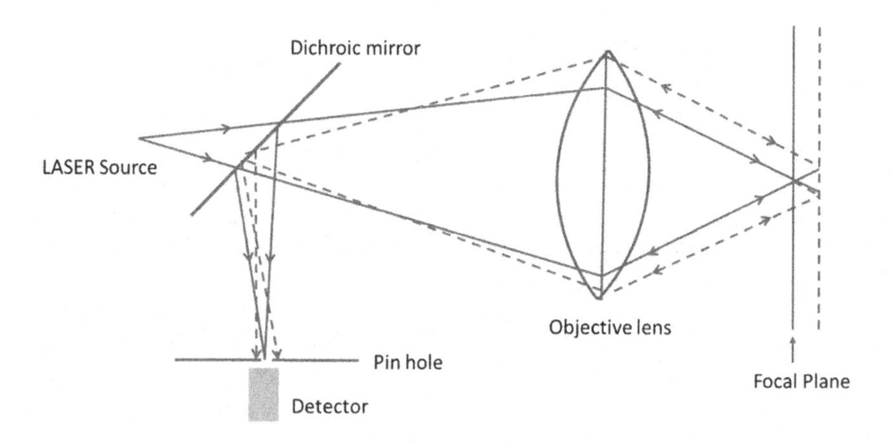

FIGURE 8.1 Working principle of confocal scanning light microscopy.

fluorescent compounds provide much more specific results as these tools will mark individual proteins and saccharides rather than whole groups of compounds (e.g. proteins). In Figure 8.1 an incident light confocal device, a diagram of the excitation beam, reflected and fluorescent light is shown. The lens functions as both a condenser and a collector. In-focus light is represented by solid lines, while out-of-focus light is represented by dotted lines.

8.3 NOVEL TECHNIQUES IN LIGHT MICROSCOPY

In light microscopy, advances in instrumentation have been developed, most notably in the advancement of confocal laser-scanning microscopy (CLSM). Despite the fact that the confocal microscope was invented in 1957, it wasn't until the 1980s that technological advances and combinations allowed for commercial development of instruments and, as a result, more widespread use of the technology. The key difference between a confocal and traditional microscope is the existence of a pinhole at the picture's focal plane, which removes out-of-focus regions, resulting in a clearer image, as well as enabling optical sectioning of the specimen, which includes focusing at predetermined levels underneath the surface. Unlike standard light microscopy, which illuminates the whole specimen or the field of view uniformly, confocal microscopy illuminates and images the specimen one point at a time through a pinhole (hence the name "scanning" microscopy). A schematic diagram demonstrating the principle of confocal laser-scanning microscopy is shown in Figure 8.2. As shown in the figure, excitation of the specimen is achieved by the emittance of light through an illumination aperture. This light passes through a dichroic mirror and is focused on the specimen through the objective lens. Fluorochrome excitation leads to the emission of a longer wavelength light, which returns via the objective and reflects off the dichroic mirror through the imaging aperture (the detection pinhole), and is detected by a photomultiplier tube. Light from above and below

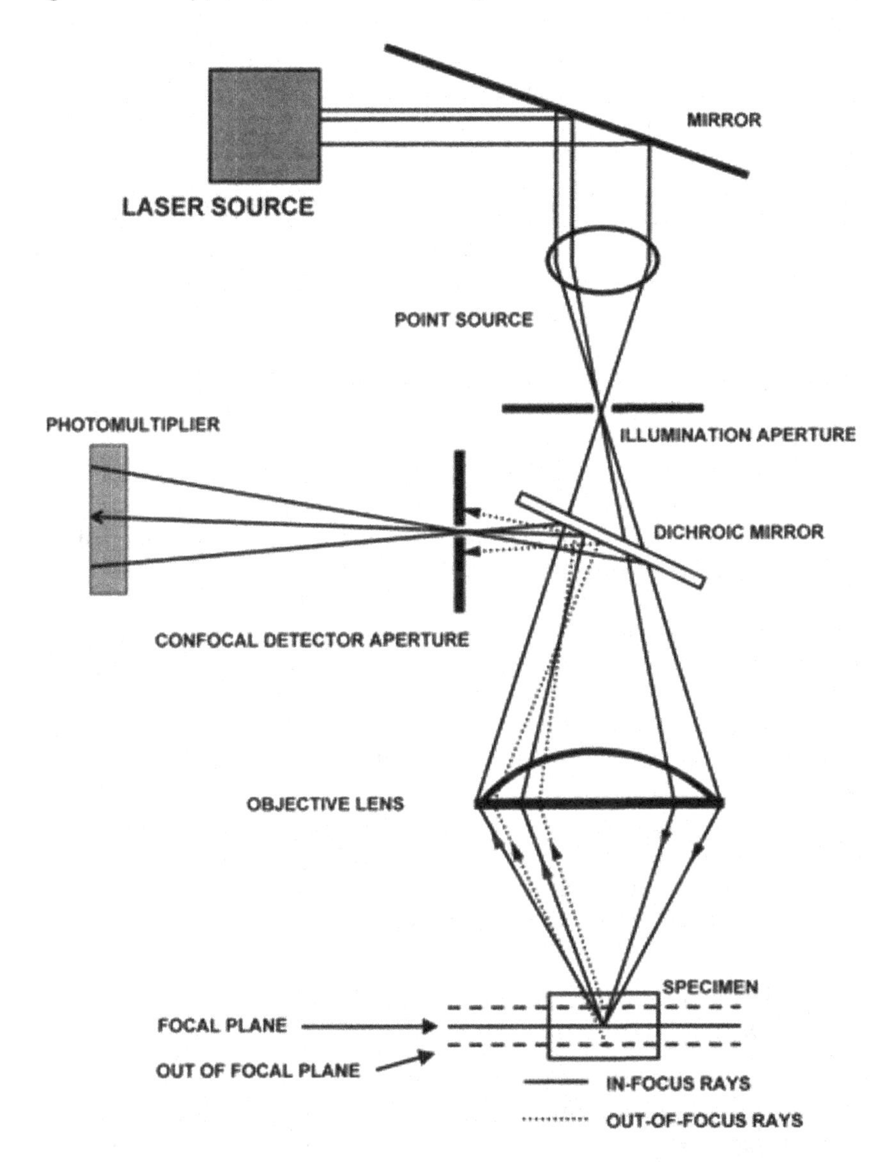

FIGURE 8.2 Working principle of confocal laser scanning microscopy [14].

the plane of focus is blocked by the pinhole. Thus, out-of-focus light is virtually eliminated from the final confocal image.

CLSM has proven most effective in the study of high-fat foods, which are difficult to prepare for traditional microscopy without losing or migrating the fat. Larger food samples may be optically sectioned using CLSM, enabling the imaging of delicate relationships within the sample that are often distorted or broken by physical sectioning or smearing techniques. CLSM has been used to describe high-fat spreads like butter and margarine, as well as lower-fat (40%) spreads like "Halvarine," by staining the lipids with Nile Red and the proteins with fluorescein isothiocyanate. The size and location of the fat globules inside the spreads are not affected by specimen preparation, so bulk specimens can be used. CLSM was also used to observe the production of the structure of wheat dough during proofing, as well as the localisation of additives in wheat gluten protein, by the same researchers. Depending on the number of laser lines available on the instrument, CLSM can recognise and localise several components at once using special fluorescent labels. Antibodies and lectins, like traditional fluorescence microscopy, significantly increase the amount of very unique protein and saccharide components that can be examined.

Another breakthrough in light microscopy is the combination of microscopy with different forms of spectroscopy, allowing particular chemical groups to be classified and mapped in situ. While much of the instrumentation is not brand new, it is only with the addition of computers to monitor the process and interpret the data that more applications have become feasible. The chemical nature of an unstained specimen can be determined down to the level of a single cell or microstructural feature using spectral characteristics. It is possible to distinguish between samples of insoluble fibre produced for use in bakery products using UV absorption, for example. Fourier transform infrared micro-spectroscopy has been used to describe the chemical essence of components in cereals, oilseeds, herbs and flavour compounds, as well as to analyse bacterial infection in potato tubers. The distribution of a specific component within a specimen can be mapped by scanning across it at a single wavelength. Fluorescence micro-spectrometry is being used to research the distribution of the polysaccharide in oat kernels, as a way of recognising appropriate varieties for processing purposes, due to interest in using mixed-linkage β-glucan from oat as a dietary adjunct to help stabilise serum cholesterol levels [13].

8.4 APPLICATION OF LIGHT MICROSCOPY IN FOOD PROCESSING

The arrangement of elements within a food, as well as their interaction, is referred to as food microstructure. During food processing, food scientists break down and reconstruct microstructure, which can be perceived as a sequence of restructuring and reassembling operations. Newer techniques for examining food microstructure and analysing images have aided researchers in this area. Since there is a causal relationship between structure and functionality, microstructural information is essential if food properties are to be regulated properly. The microstructural approach to food processing and engineering is focused on the realisation that product properties are influenced by microstructural shifts and that microstructural techniques are needed to comprehend structure–property relationships.

LM is a technique for determining the morphology of cells and tissues in biological materials. This technique's capacity for analysing the structure of food products is also recognised, but it is not well known yet. Similarly, information on LM studies of starch's supramolecular structure is particularly scarce. Light microscopy also has a lot of scope for starch research, thanks to recent advancements in microscopic techniques, such as increased resolution and improved image processing methods. The ability to distinguish between amylose and amylopectin using iodine staining distinguishes LM from other microscopic techniques. As a result, LM is particularly useful for studying the gelatinisation of starch, the degree of molecular dispersion of its macromolecules and the structural changes induced by alteration. Furthermore, it can be especially useful for analysing changes in the supramolecular structure of starch in a food product matrix, providing more detail than SEM, which is the most commonly used technique.

Food components such as lipids and proteins can be stained selectively prior to processing, or the stain can be diffused into the product. Lipids and proteins may be identified with unique stains. These stains can be adsorbed to the microstructure of interest or can be covalently coupled to it. Antibodies raised against specific proteins can be combined with fluorescent markers that can be identified by CLSM, allowing for highly selective labelling of specific proteins. To mark carbohydrates and proteins with carbohydrate residues, a technique based on the basic affinity of lectins for carbohydrates can be used.

Although advanced microscopic techniques such as AFM (Atomic Force Microscope), CLSM or SEM provide more comprehensive details about the morphology of the starch granule, LM can be used to illustrate structural changes in dispersed systems. Iodine staining will quickly differentiate amylose from amylopectin when LM is used. Furthermore, under LM, the degree of molecular dispersion of starch in pastes, as well as changes in the polymer–solvent interaction induced by the replacement of hydroxyl groups with polar or non-polar moieties, can be investigated (Table 8.1).

8.5 CONCLUSION

Food structure imaging is expected to follow in the footsteps of advances in biology, medicine and materials science. Trends in biology are oriented towards a wider variety of applications, with the ultimate goal of improving image quality and contrast, expanding the range of specimen forms that can be imaged, and reducing the risk of drawing incorrect conclusions from images by improving image analysis techniques. Although the structure of certain foods (such as milk, poultry, cereal products and legumes) has been extensively studied, the structures of many others have yet to be investigated. The structure of bakery items made from different grains, cheeses made from goat, sheep or cow's milk and roast pork from animals fed various diets is being studied. Processes such as milling, comminuting, heating and extrusion can be explored further. It is no longer sufficient to classify a food by comparing it to other micrographs using so-called representative micrographs and qualitative visual evaluation. Microscopy is increasingly becoming objectified, with images being converted

TABLE 8.1
Applications of different imaging techniques in food industry

Microscopic techniques	Application	Image (Example)
Light microscope	To determine the morphology of cells and tissues	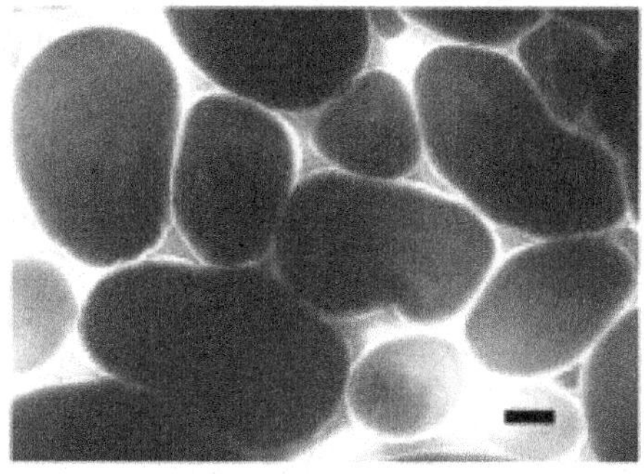 Fig. Distarch of potato origin [15]
CLSM	To obtain high-resolution images with all areas in focus	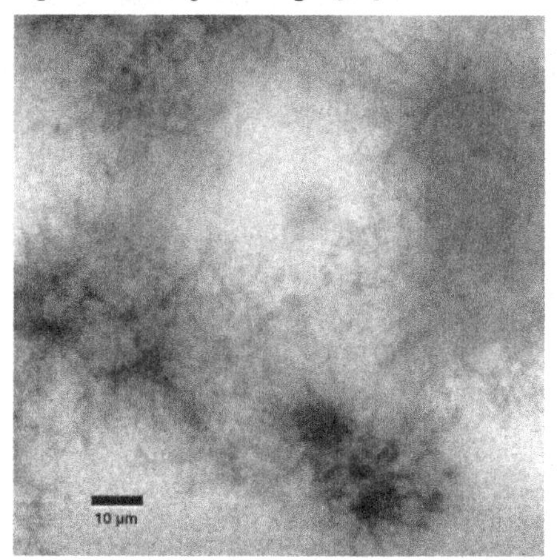 Fig. Confocal laser scanning micrograph of milk fat [16]
AFM	To identify the type of surface and mechanical properties	

TABLE 8.1 *(Continued)*
Applications of different imaging techniques in food industry

Microscopic techniques	Application	Image (Example)
		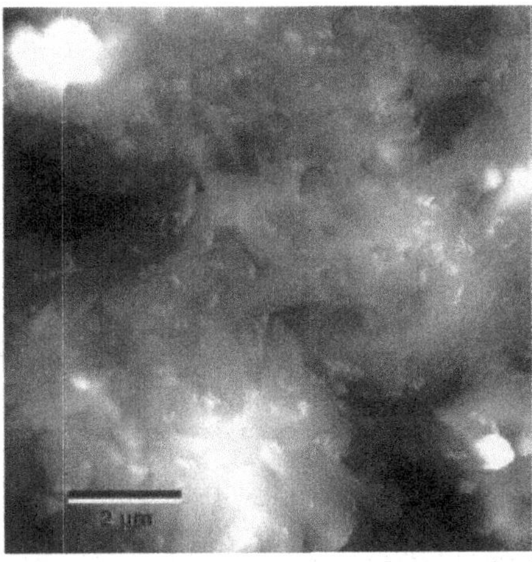 Fig. Tapping mode atomic force microscopy of palm oil [16]
SEM	To determine the surface changes	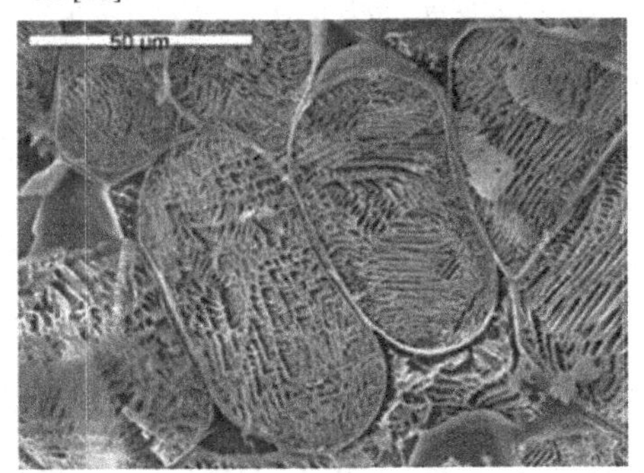 Fig. Natesa carrot cells obtained by SEM [17]

into numerical data. Digital images would almost certainly be subjected to image and fractal analysis on a regular basis.

Structural studies are expected to play a larger role in elucidating the relationships between sensory qualities of foods, such as texture, and food structure, which will have a greater effect on the development of new foods. Structural studies will almost certainly become an integral part of biotechnology research, as they will be used to mask structural changes in foods created by genetic manipulation. Finally, image analysis is expected to become increasingly important in the food industry, especially in quality control and as a robot vision component of the production line.

REFERENCES

[1]. Rahman, M.M., C. Kumar, M.U. Joardder, and M. Karim, *A micro-level transport model for plant-based food materials during drying,* Chemical Engineering Science, 2018. **187**: pp. 1–15.

[2]. Joardder, M.U., C. Kumar, R.J. Brown, and M. Karim, *A micro-level investigation of the solid displacement method for porosity determination of dried food,* Journal of Food Engineering, 2015. **166**: pp. 156–164.

[3]. Rahman, M.M., M.U. Joardder, and A. Karim, *Non-destructive investigation of cellular level moisture distribution and morphological changes during drying of a plant-based food material,* Biosystems Engineering, 2018. **169**: pp. 126–138.

[4]. Rahman, M.M., M.U. Joardder, M.I.H. Khan, N.D. Pham, and M. Karim, *Multi-scale model of food drying: Current status and challenges,* Critical Reviews in Food Science and Nutrition, 2018. **58**(5): pp. 858–876.

[5]. Duc Pham, N. *et al.*, *Quality of plant-based food materials and its prediction during intermittent drying,* Critical Reviews in Food Science and Nutrition, 2019. **59**(8): pp. 1197–1211.

[6]. Kumar, C., M.U. Joardder, T.W. Farrell, G.J. Millar, and A. Karim, *A porous media transport model for apple drying,* Biosystems Engineering, 2018. **176**: pp. 12–25.

[7]. Khan, M.I.H. and M. Karim, *Cellular water distribution, transport, and its investigation methods for plant-based food material,* Food Research International, 2017. **99**: pp. 1–14.

[8]. Khan, M.I.H., T. Farrell, S. Nagy, and M. Karim, *Fundamental understanding of cellular water transport process in bio-food material during drying,* Scientific Reports, 2018. **8**(1): p. 15191.

[9]. Kaláb, M., P. Allan-Wojtas, and S.S. Miller, *Microscopy and other imaging techniques in food structure analysis,* Trends in Food Science & Technology, 1995. **6**(6): pp. 177–186.

[10]. Abesinghe, A., N. Islam, J. Vidanarachchi, S. Prakash, K. Silva, and M. Karim, *Effects of ultrasound on the fermentation profile of fermented milk products incorporated with lactic acid bacteria,* International Dairy Journal, 2019. **90**: pp. 1–14.

[11]. Svegmark, K. and A.-M. Hermansson, *Distribution of amylose and amylopectin in potato starch pastes: Effects of heating and shearing,* Food Structure, 1991. **10**(2): p. 2.

[12]. Aguilera, J. and D. Stanley, *Microstructural Principles of Food & Engineering,* 1990. Elsevier.

[13]. Blonk, J. and H. Van Aalst, *Confocal scanning light microscopy in food research,* Food Research International, 1993. **26**(4): pp. 297–311.

[14]. Mahmudi-Azer, S., P. Lacy, and R. Moqbel, Tracing Intracellular Mediator Storage and Mobilization in Eosinophils. In: Rogers, D.F., Donnelly, L.E.(eds). Human Airway Inflammation. Methods in Molecular Medicine, 2001. Humana Press. https://doi.org/10.1385/1-59259-151-5:367

[15]. Błaszczak, W. and G. Lewandowicz, *Light Microscopy as a Tool to Evaluate the Functionality of Starch in Food,* Foods, 2020. **9**(5): p. 670.

[16]. Aguilera, J.M., D.W. Stanley, and K.W. Baker, *New dimensions in microstructure of food products*, Trends in Food Science & Technology, 2000. **11**(1): pp. 3–9.
[17]. Aguilera, J.M., *Drying and dried products under the microscope,* Food Science and Technology International, 2003. **9**(3): pp. 137–143.

9 Neutron Radiography Scattering in Food Processing

9.1 INTRODUCTION

Food processing, such as drying, helps food preservation, facilitates the easier handling of foods and destructs microorganisms in the food [1]. The quality of the processed food can be deteriorated due to chemical and physical changes during processing [2]. These changes can also alter the texture and the food structure. Neutron radiography is one of the most prominent technologies for the micro-level investigation of food materials during processing [3].

Like X-ray imaging, neutron radiography is a micro-level non-destructive imaging technique [4]. An image is formed when a neutron beam is directed towards the sample. Some structural elements in the sample which cannot be seen in the X-ray image are visible in neutron radiography. The primary application of neutron imaging is to investigate the microstructure of the materials. This technique is more advantageous than other imaging techniques in obtaining the images of hydrogen or other specific light nuclides of water transport [5].

Neutron radiography is mainly used for the quality control inspection of manufactured components in the aircraft, aerospace and automotive industries as they require precision machining [6,7]. Real-time imaging also allows fast movement (few seconds) of the objects. This technology has also been used to monitor the boron distribution required for cancer treatment [5]. Neutron radiography has potential application in the food science and biomedical fields as well [3,5,8]. For example, it has been used to investigate the water movement in plants, leaves and flowers, including the water distribution and transport mechanisms [3,9].

Neutron imaging obtains not only the water content in food materials but also the spatial distribution. Neutron radiography can accurately investigate the moisture and gas transport in food materials during processing due to its high sensitivity [10]. The density of food materials is one of the essential characteristics for assessing processed food quality, which also indicates the rheological and mechanical properties of the processed food [11]. Density can easily be measured by the neutron radiography method. This technology effectively provides very high-resolution data (spatial, temporal and dynamic) on internal water transport during food processing. These data are beneficial for formulating and validating food processing models.

DOI: 10.1201/9781003047018-9

9.2 NEUTRON AND X-RAY TOMOGRAPHY

Neutron tomography is a cutting-edge method for determining local moisture distribution, and neutron imaging works on the same basis as X-ray imaging [12]. Both investigate the attenuation of the incoming radiation (neutrons in the former, photons in the latter) and how it relates to the material's local structure. The attenuation of X-ray radiation is proportional to the atomic number since it interacts with the outer electron shells of the atoms it meets (and their density). On the other hand, in neutron scattering, neutrons interfere with the nuclei of the atoms [13]. The high attenuation of hydrogen atoms makes it particularly useful for determining the moisture content (and thus any drying fronts) in a material.

An integrated neutron and X-ray imaging equipment was developed to take the advantages of both systems [12]. The system is shown in Figure 9.1, which compares the horizontal slices of X-ray and neutron topographies. Figure 9.2 shows an example of a 1-minute tomography of a concrete sample subjected to a unidirectional 500°C heating in situ, highlighting in white the receding moisture profile and in black the aggregates, whose detection is simplified by the combined use of X-ray and neutrons. Because of their similar density and atomic number, X-rays are suitable for detecting pores in concrete (due to their low density), but they cannot differentiate mortar and quartz grains used as aggregates. The quartz aggregates, on the other hand, are almost transparent to neutrons, making it easy to distinguish them from the mortar but difficult to

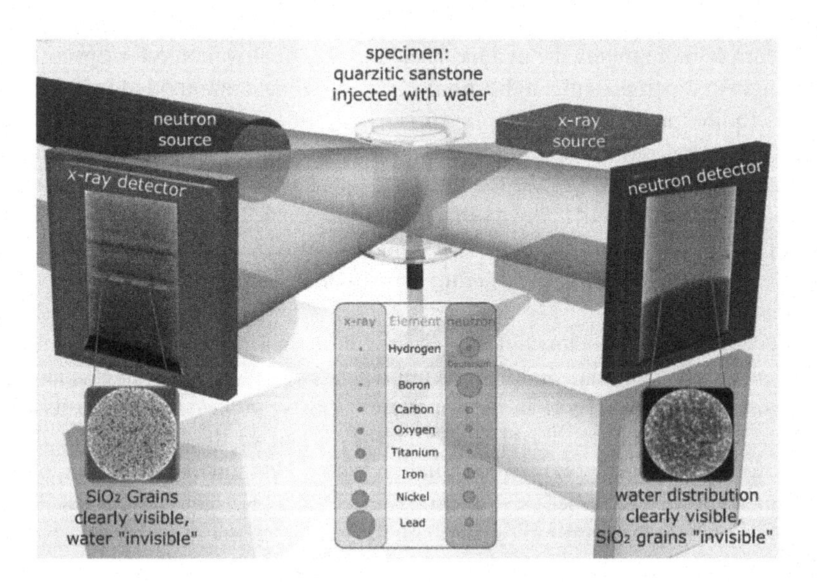

FIGURE 9.1 X-ray is best for detecting and measuring the solid skeleton, while neutrons are best for studying the hydraulic response [12].

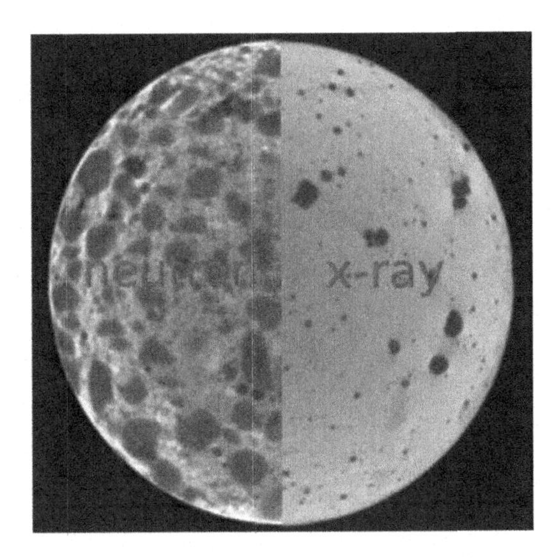

FIGURE 9.2 The same cylindrical concrete sample was subjected to neutron and X-ray tomography [12].

distinguish them from the pores. This enables researchers to monitor the evolution of the cement paste's moisture content.

It can be noted that tomography efficiency (in terms of spatial resolution, temporal resolution and measurement of noise) is entirely dependent on noise in the collected data [4]. Another important aspect of the combined instrument is the ability to conduct X-ray absorption imaging, which takes advantage of the high complementarity of these two techniques, as seen in Figure 9.2. As illustrated in several articles, neutron imaging has a wide variety of applications ranging from green energy to biology and palaeontology to porous media.

9.3 APPLICATION OF NEUTRON IMAGING IN FOOD PROCESSING

A multiphysics model for coupled moisture transport and mechanical deformation, as well as quantitative neutron tomography, was used to investigate the dehydration of cylindrical apple tissue [10]. This is the first time that the dynamics of water distribution and shrinkage (deformation) of food tissue during drying has been analysed in 3D using both neutron tomography and computational modelling [10]. The heterogeneity of the water distribution in apple tissue was discovered using neutron tomography. As compared to gravimetric measurements, the neutron experiments can accurately estimate the water loss from the samples, suggesting that this method accurately quantifies water content for this type of experiment. It is evident that neutron tomography has the ability to monitor the dehydration of fruit and vegetable samples (Figure 9.3).

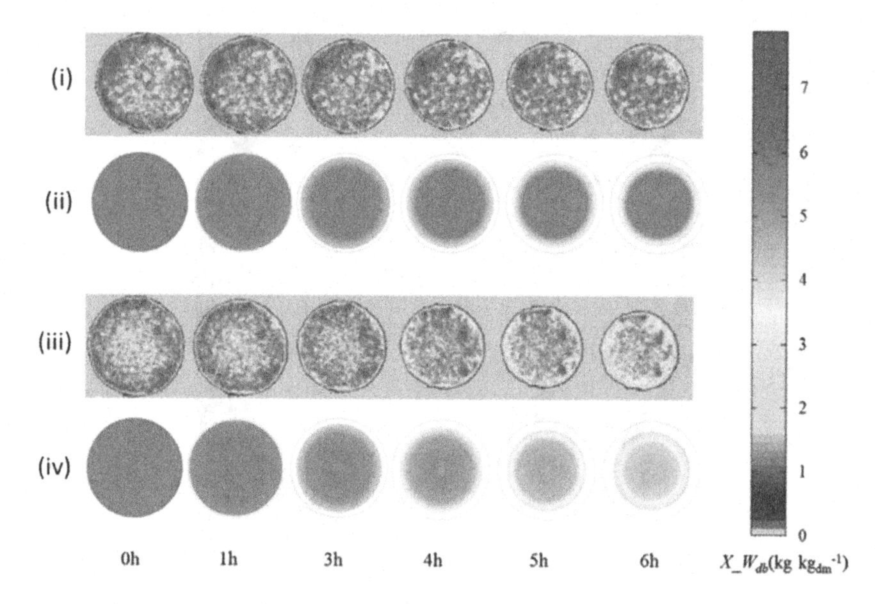

FIGURE 9.3 Dehydration of water through the apple tissue as a function of drying time. (i) Tomography image (y = 10 mm). (ii) Results of the numerical model (y = 10 mm). (iii) Tomography image (y = 17.5 mm), (iv) Results of the numerical model (y = 17.5 mm) [10].

Figure 9.4 shows the neutron radiography test set up for the convective drying of food. The setup includes a detector with a CCD camera system to capture the overall phenomena during processing. The camera has the capability to take pictures up to 83 µm. This water content can be determined gravimetrically as the total amount of water in the fruit sample to the initial sample volume. The exposure time of the radiography was 8 second.

The water loss of the food materials can be quantified by intensity measurements of a transmitted neutron beam. The Beer–Lambert law can describe the intensity measurement [10]:

$$I(t) = I_0 e^{-\mu . z} \tag{9.1}$$

where $I(t)$ is the transmitted monochromatic beam at a specific time, I_0 is the intensity of the incident neutron beam, μ is the effective attenuation coefficient for neutrons and z is the thickness of the object along the beam direction.

The neutron radiograph needs correction before the quantitative analysis. The standard correction includes dark current correction, intensity correction, flat field correction, black and body correction. The dark current modification is mainly removing the background noise of the image taken by the CCD camera. Intensity correction refers to averaging the beam

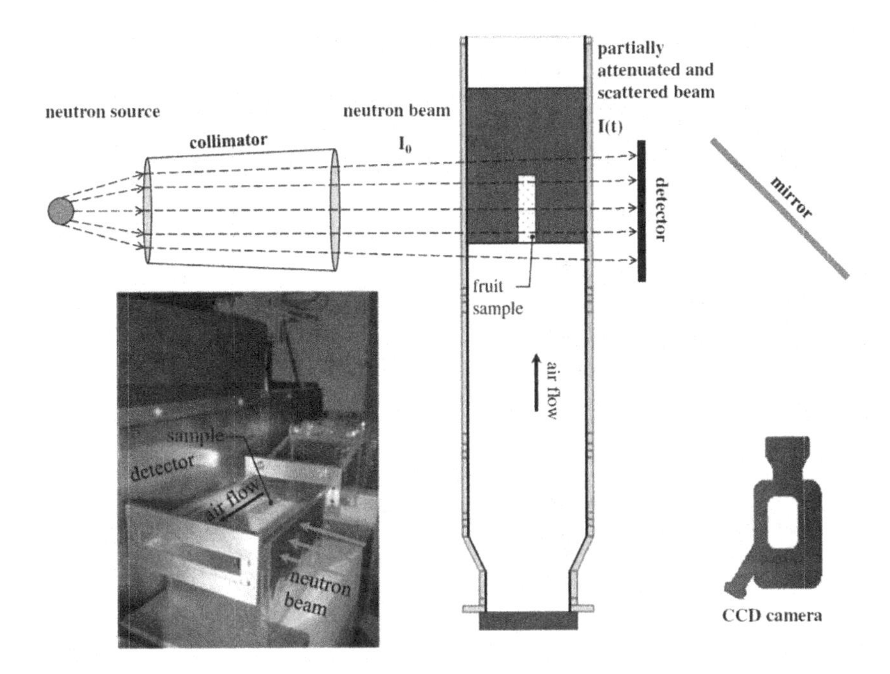

FIGURE 9.4 Schematic diagram of neutron radiography experiment in food processing [14].

variation if there is any fluctuation. Flat field correction eliminates the spatial inhomogeneities from the images, performed by correcting pixels in each image. Black body correction refers to removing the scattering neutron signal from scattering by the overall experimental configuration by subtracting a constant value from the pictures. A special algorithm named quantitative neutron imaging (QNI) was developed to perform all the corrections [10].

Here, the neutron tomography investigation of two fruits (apple and pears) during processing (convective drying) has been presented as examples of neutron radiography application in food processing. Internal moisture distribution of apple and pears at different times of drying captured by the neutron tomography is presented in Figure 9.5. The evolution of the moisture distribution and the water loss pattern is understandable from the neutron tomography images. It can be seen from the photos that the water loss process in both the apple and the pear is similar. However, the apple dries faster than the pears. It is also visible that the pears lose water faster at the beginning of the process. The variation of the water loss in both fruits happens due to the internal water transport properties. The drying rate is every species and sample size dependent. Moreover, the diffusion coefficients are dependent on the shelf life and temperature.

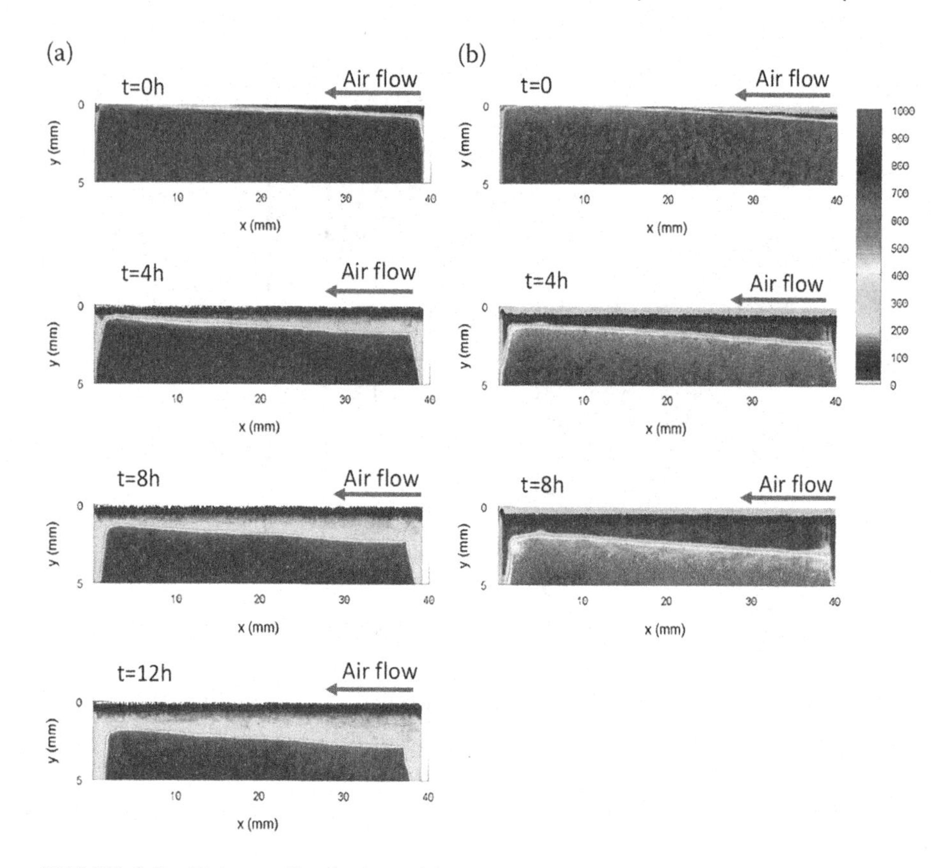

FIGURE 9.5 Moisture distribution of (a) pears and (b) apple obtained by neutron tomography during drying [10].

9.4 CONCLUSION

This chapter briefly presents the neutron radiography method in a food processor, along with some experimental results. The application has been explained by using the food drying process as an example, and water content quantification and distribution in fruit and vegetables has been presented. Food drying involves simultaneous heat, mass and momentum transfer with continuous phase change. The technique is prevalent in providing quantitative data of the internal water distribution inside fruits and the water loss during drying. The major benefit of this method is that the data can be obtained in quasi-real time at a high dynamic resolution and high spatial. This technique has been essential for studying the processing of food non-destructively. It can provide detailed data for validating a food processing model, including the moisture transport information.

REFERENCES

[1]. Joardder, M.U.H., C. Kumar, and M.A. Karim, *Food structure: Its formation and relationships with other properties.* Critical Reviews in Food Science and Nutrition, 2017. **57**(6): pp. 1190–1205.

[2]. Khan, M.I.H. and M. Karim, *Cellular water distribution, transport, and its investigation methods for plant-based food material.* Food Research International, 2017. **99**: pp. 1–14.

[3]. Esser, H.G., et al., *Neutron radiography and tomography of water distribution in the root zone.* Journal of Plant Nutrition and Soil Science, 2010. **173**(5): pp. 757–764.

[4]. Cleveland, T.E., et al., *The use of neutron tomography for the structural analysis of corn kernels.* Journal of cereal science, 2008. **48**(2): pp. 517–525.

[5]. Loupiac, C., et al. *Neutron imaging and tomography: Applications in food science.* In *EPJ Web of Conferences.* 2018, EDP Sciences.

[6]. Bilheux, H.Z., R. McGreevy, and I.S. Anderson, *Neutron imaging and applications: A reference for the imaging community.* 2009, Springer.

[7]. Wang, Y., et al., *Nuclear instruments and methods in physics research section B: beamBeam interactions with materials and atoms.* Nuclear Instruments and Methods in Physics Research B, 2001. **180**(1-4): pp. 251–256.

[8]. Lehmann, E., S. Hartmann, and P. Wyer, *Neutron radiography as visualization and quantification method for conservation measures of wood firmness enhancement.* Nuclear Instruments and Methods in Physics Research Section A: Accelerators, Spectrometers, Detectors and Associated Equipment, 2005. **542**(1-3): pp. 87–94.

[9]. Matsushima, U., et al., *Estimation of water flow velocity in small plants using cold neutron imaging with D2O tracer.* Nuclear Instruments and Methods in Physics Research Section A: Accelerators, Spectrometers, Detectors and Associated Equipment, 2009. **605**(1-2): pp. 146–149.

[10]. Aregawi, W., et al., *Dehydration of apple tissue: intercomparison of neutron tomography with numerical modelling.* International Journal of Heat and Mass Transfer, 2013. **67**: pp. 173–182.

[11]. Defraeye, T., et al., *Novel application of neutron radiography to forced convective drying of fruit tissue.* Food and Bioprocess Technology, 2013. **6**(12): pp. 3353–3367.

[12]. Tengattini, A., et al., *NeXT-Grenoble, the neutron and X-ray tomograph in Grenoble.* Nuclear Instruments and Methods in Physics Research Section A: Accelerators, Spectrometers, Detectors and Associated Equipment, 2020. **968**: p. 163939.

[13]. Dauti, D., et al., *Analysis of moisture migration in concrete at high temperature through in-situ neutron tomography.* Cement and Concrete Research, 2018. **111**: pp. 41–55.

[14]. Aregawi, W., et al., *Understanding forced convective drying of apple tissue: Combining neutron radiography and numerical modelling.* Innovative Food Science & Emerging Technologies, 2014. **24**: pp. 97–105.

Index

Printed in the United States
by Baker & Taylor Publisher Services